Introduction to Nuclear Power

Series in Chemical and Mechanical Engineering

G.F. Hewitt and C.L. Tien, *Editors*

Introduction to Nuclear Power

Second Edition

Geoffrey F. Hewitt

Department of Chemical Engineering
and Chemical Technology
Imperial College
London, U.K.

John G. Collier*

*Deceased
Formerly Chairman
Nuclear Electric plc
Barwood, U.K.

CRC Press
Taylor & Francis Group
Boca Raton London New York

CRC Press is an imprint of the
Taylor & Francis Group, an **informa** business
A TAYLOR & FRANCIS BOOK

USA	Publishing Office	Taylor & Francis
		29 West 35th Street
		New York, NY 10001-2299
		Tel: (212) 216-7800
	Distribution Center	Taylor & Francis
		47 Runway Road, Suite G
		Levittown, PA 19057-4700
		TEL: (215) 269-0400
		FAX: (215) 269-0363
UK		Taylor & Francis
		11 New Fetter Lane
		London EC4P 4EE
		Tel: 011 44 207 583 9855
		Fax: 011 44 207 842 2298

INTRODUCTION TO NUCLEAR POWER

A CIP catalog record for this book is available from the British Library.

Library of Congress Cataloging-in-Publication Data

Collier, John G. (John Gordon), 1935–
Introduction to nuclear power/John G. Collier, Geoffrey F. Hewitt. —2nd ed.

   p.   cm.
Includes bibliographical references and index.
ISBN 1—56032—454—6 (hardcover: alk. paper)
1. Nuclear energy.   2. Nuclear power plants.
I. Hewitt, G. F. (Geoffrey Frederick).   II. title.

TK9145.C584 1997
621.48'3—dc21
                                                 97-13559
                                                 CIP

Contents

Preface
To the Second Edition

This Second Edition has been several years in the making. My life-long friend and colleague John Collier died from pancreatic cancer on November 18, 1995. This Second Edition must sadly but proudly serve as a memorial to John and to his intense and firm conviction of the need for nuclear power for the future well-being of the human race on this planet. John Collier's transparent honesty and humanity provided the best possible witness to the sincerity of this conviction. I, too, strongly believe in the ultimate necessity for nuclear power; there will be temporary situations where this need is not so obvious (for instance, the current availability of an excess of natural gas in the United Kingdom), but the long-term situation is clear. It is thus vital to continue research and development in the area and to maintain an adequate technology base. Everything possible must be done to develop public confidence in nuclear power, and the industry should not be averse to considering new concepts which spring from the lessons regarding inherent safety learned in the chemical industry. The main public concern is with the possibility of severe accidents, and the accidents at Three-Mile Island and Chernobyl have naturally served to fuel this fear. The nuclear industry must recognize this problem of public acceptability and face up to it. Once the long term need for nuclear power is recognised and accepted, solutions can and indeed must be found. However, it is worth pointing out that of all modern industrial plant, even the present generation of nuclear power stations is among the safest. In a properly regulated environment, the present operating nuclear power stations provide a safe and economic means of energy production. However, the nuclear industry needs to give a lot more thought to the sources and consequences of major accidents if, as it seems inevitable to me, nuclear power generation will need to be expanded to meet the growing energy demands. It is with this as a background that a large amount of the material in this book is concerned with nuclear accidents and their consequences.

For this Second Edition, the material has been extensively updated and revised. In the months before his death, John Collier carried out much of the work in preparation for this, and I would like to place on record my appreciation of his contribution. Perhaps the most important new material is that associated with the Chernobyl accident. This accident happened on April 28, 1986, at a time when the proofs of the First Edition had been produced. A short section was written in the First Edition about the accident but, of course, a full realization of the sources and consequences of the event was not at that stage possible. We have attempted to rectify this in the current volume. We have also updated the section on the Three Mile Island accident to reflect the continuing developments in understanding and analysis of that event.

Other major modifications in the current volume, with respect to the First Edition, include an updating of the material on Earth's internal heat generation in Chapter 1, major updating and revision of the general material on severe accidents, and an updating of the material relating to fusion power generation.

I hope that this new edition will be a helpful update for those who purchased and used the First Edition and that it will serve to introduce a new generation of readers to nuclear power and its enormous future potential.

G.F. Hewitt, 1996

Preface
To the First Edition

The decision to write this book was made several years ago against a background of general unease that we both felt about the level of public understanding of nuclear power and its associated technologies. There is no doubt that there are currently considerable fears in the minds of many people about nuclear power generation. Unless these fears are dispelled through a deeper and more widespread understanding of the technologies and other issues involved, the development of nuclear power, which has a vital contribution to make to the world's energy requirements, may be jeopardized.

In preparing this book we have tried our utmost to present nuclear power in simple terms as it really is. Thus, we have discussed real and actual accident scenarios in detail, just as we have discussed the problems of disposal of nuclear waste. Our aim has been to give a factual and unemotional presentation of what is now a relatively mature technology. This book was in production when news of the Chernobyl reactor accident in the USSR emerged. We have included some material on this reactor type and, as best as we can, the information available about the accident itself. The worldwide concern following this accident has illustrated again very directly the need for better and simpler information to be available to the public about nuclear power.

One of the major difficulties in writing a general introductory book of this kind is that of deciding the level and type of audience to which it should be addressed. Our overall aim has been to produce a text that is as free of jargon as is possible and that demands the minimum possible basic scientific knowledge, while at the same time presenting descriptions and facts at a level of detail sufficient to make them generally useful. Thus, the text should be of interest to a variety of readers, including the following:

1. The intelligent general reader, interested in science and technology, who wishes to brief him or herself in greater depth about nuclear power.
2. The undergraduate or graduate student pursuing introductory courses on energy in general and nuclear power in particular. It was with this student audience in mind that we have given a number of worked examples and problems at the end of each chapter; these are designed to increase the depth of understanding of the concepts described and to provide an aid to the use of the text in presenting such courses.
3. The industrial technologist wishing to obtain an overview of the nuclear industry. It is perhaps typical of the pressures of modern life that many technologists, even within the nuclear industry itself, do not have a full general appreciation of the overall basis of nuclear power. This book should, we hope, help fill that gap.

Both of us were trained as chemical engineers (JGC at University College, London, and GFH at UMIST, Manchester), and we have both specialized in the thermal aspects of nuclear power. It is from this viewpoint that the book has mainly been written. We make no apologies for this; the generation and dissipation of heat have a dominant position in nuclear power. Heat generation is important not only during the time of operation of the nuclear reactor but also in considering what happens to the nuclear fuel once it is removed from the reactor. Because of the fission products, heat generation continues at a significant rate for decades after the fuel is taken out of the reactor. Careful consideration must, therefore, be given to cooling the fuel at all stages, and this will be the theme that forms a consistent thread throughout the book.

We gratefully acknowledge the considerable assistance we had from a number of people in preparing the final manuscript. In particular, we thank Sonya Crowe and Mary Phillips Born, who read the manuscript from the nonspecialist viewpoint. They, and several other readers, helped us identify unnecessary jargon in the original manuscript and pinpoint parts of the text where the explanations were less clear than they ought to be. We are also very grateful to our colleagues at Harwell and in the CEGB for assistance in the preparation of the diagrams, checking of the examples, and typing and preparing the manuscript, although we stress that any views and opinions are our own. Finally, we would like to thank our wives (Ellen and Shirley) for their support and patience. Despite their good efforts to keep us apart, we fear that (by continuing our incessant conversations on nuclear power and two phase heat transfer) we have not given them the support that we should at many a cocktail and dinner party!

The objective of this book is to introduce nuclear power in a factual and unemotional manner. However, in all fairness to the reader, we must close this preface by stating our own position quite unequivocally. Notwithstanding fluctuations caused by recessions, supply difficulties, oil price rises and slumps, etc., there is a continuous underlying increase in humanity's demand for energy. This will continue and accelerate as the underdeveloped countries begin to demand standards that we now take for granted in the industrialized nations. The fossil fuels (coal, gas, and oil) are finite and, as we all realize, recovery may ultimately prove uneconomic or their use unacceptable as the demand for global environmental protection grows. Alternative energy sources (tidal, solar, geothermal, and wind) all have their place and deserve continuing support and development; however, even the most optimistic of their proponents, cannot see them becoming the major component of the growing bulk energy requirement. Energy conservation, too, is vitally important and must be encouraged with the maximum attention. However, neither alternative sources nor energy conservation is likely to bridge the gap between demand and supply over the next century, and nuclear power is the important and growing energy source for the future. It is a clean and efficient power source, both economic and compact, with a minimum of environmental impact. Accidents like Three Mile Island and Chernobyl need to be put firmly into context with other industrial accidents and particularly those related to the energy industry. However, like any other technology, nuclear power must be developed responsibly and the facts about it clearly understood and accepted by the public and also by those in government who make decisions on technology policy. That is why we wrote this book.

John G. Collier
Geoffrey F. Hewitt

1

The Earth and Nuclear Power
Sources and Resources

1.1 INTRODUCTION

This book is written from an engineer's viewpoint, particularly that of a thermal engineer, that is, a design or research engineer concerned with heat production and utilization. We believe that the most important problems in the utilization of nuclear power concern the handling of thermal energy generated in the various processes. This includes handling under the normal operating and processing conditions and dealing with heat removal problems under the unlikely conditions of an accident. The problem of handling thermal energy associated with nuclear power does not stop when the fuel is removed from the power station; small amounts of heat are generated in the spent fuel before it is processed and in the waste products. The consequences of this are also the concern of the thermal engineer.

The approach that we shall take, therefore, is one that is not normally followed in general books on nuclear energy. We will follow the history of nuclear materials from their cosmic origins, through their terrestrial life span up to the time when they are used in nuclear reactors, and beyond. Although we will need to explain some elementary aspects of physics, the emphasis will be on what happens to the thermal energy.

We begin with the history of uranium in the earth, the decay of its isotopes, and the effect this decay has had on the earth as we know it. Comparisons are made with the earth's other main energy source: the sun. Energy from the sun is derived either directly or through storage media such as fossil fuels, hydro-electric power, and winds.

The rate at which energy may be extracted from nuclear materials can be enhanced by the self-sustaining process of nuclear fission. Nuclear fission does not normally occur in nature, but recent studies have revealed that nature anticipated Enrico Fermi by about 2 billion years in creating a natural nuclear fission reactor by a series of extraordinary and improbable events. We shall use this example in introducing nuclear fission.

In the final part of this chapter, we compare the relative magnitudes of thermal energy resources of the various types: fossil fuel, nuclear, solar, and so forth.

1.1.1　Forms of Energy

What is energy? There is general awareness of the problem of depletion of the world's energy resources. People understand energy in terms of those resources, namely, the supplies of oil, gas, and coal and the electricity derived from them. All of these items have made an increasingly large demand on national and personal budgets.

The engineer has, by training, a somewhat different concept of energy. This derives from his or her undergraduate training in the field of *thermodynamics*, which is the science of energy and energy conversion. We do not intend to try to provide a basic course in thermodynamics; however, for the rest of this book to be reasonably intelligible, it is important that some of the basic concepts be stated.

The concept of doing *work* to lift objects or to move an object such as a bicycle along is a commonly accepted one. Thus, it is relatively easy to understand the concept of *energy* as a measure of the ability to do work. Energy can appear in different forms as follows:

1. *Kinetic Energy*. This is energy associated with movement, for example, that of a flywheel or a moving locomotive.

2. *Potential Energy*. This is energy possessed by virtue of position, typically in the earth's gravitational field. For instance, a child sitting on the higher end of a seesaw has greater potential energy than a child sitting on the lower end. Likewise, water in a mountain lake has greater potential energy than water at sea level.

3. *Chemical Energy*. Matter consists of atoms that are combined together in molecules. Molecules of different substances can react to release energy,

and this releasable energy is often termed *chemical energy*. For example, chemical energy is released when gasoline combines with air in the cylinders of a car's engine.

4. ***Electrical Energy***. Atoms consist of a central mass, known as the nucleus, around which a cloud of *electrons* circulates (see Figure 1.1). If there is an excess or deficit of electrons in one part of a body, the body is said to have an electrical charge and, by virtue of this, to have *electrical energy*. An example of this is a thunderstorm, where the clouds are charged electrically with respect to the ground.

5. ***Nuclear Energy***. Normally, the nucleus of an atom is stable and will remain indefinitely in its present state. An example is the nucleus of an atom of iron; no matter how much we would like it to happen, iron will never change into another element, such as gold. However, the atoms of some elements are unstable and can change into another form spontaneously, by the emission of radiation. We shall discuss the forms of radiation emitted further in Section 1.2; it is sufficient here to note that the radiation emitted has kinetic energy and the disintegration process results in the release of energy associated with the nucleus, namely, the *nuclear energy*. If the nucleus could be weighed before the disintegration, and the resulting nucleus and all particulate components of the radiation weighed afterward, it would be observed that a small change in mass had occurred due to the conver-

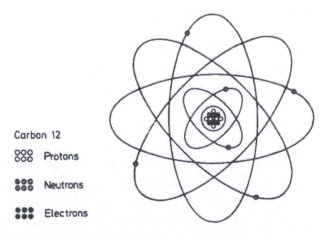

Carbon 12

Protons

Neutrons

Electrons

Figure 1.1: Schematic diagram of carbon-12 atom.

sion of mass into energy. The relationship between the loss of mass m and the energy released E is given by Einstein's famous equation:

$$E = mc^2$$

where c is the velocity of light, namely 300,000 kilometers per second (186,000 miles per second). The amount of energy deriving from a mass loss is enormous; for example, 100 kilograms of mass fully converted into energy would supply all the energy needs of the United Kingdom (at the present rate of usage) for a year. Each kilogram of mass, fully converted, is equivalent to the energy available by burning 3 million tons of coal. In a typical nuclear reaction, however, only a tiny fraction of the mass is converted into energy, typically ~0. 1%. The disintegration of an unstable nucleus, and the consequent release of nuclear energy, can be stimulated by exciting the nucleus by bombarding it with radiation. This is at the heart of the fission reaction process, which we shall discuss further below. Nuclear energy can also be released, as we shall see, by the *fusion* of very light atoms into heavier ones.

6. *Thermal Energy.* The atoms of all substances are in constant motion. In a solid the atoms are held in an approximately fixed position with respect to one another. However, they all vibrate to an extent that increases with increasing temperature. The energy associated with this vibration is called *thermal energy.* In *fluids* (namely, *liquids and gases*), two or more atoms may be combined with each other chemically in the form of *molecules.* These molecules have vibrational energy, but in the fluid state they may also have *translational energy* arising from their motion in space and *rotational energy* arising from their rotation. All of these components of energy add up to the *thermal energy* of the fluid. It will be seen from this description that thermal energy is of a special type. It is associated with atomic or molecular movements that are randomly directed. This makes it very much more difficult to convert thermal energy into other forms of energy, as we shall see below.

The intensity of atomic or molecular movement is a measure of the energy content of a piece of matter. A body that has a high intensity of atomic or molecular movement will transfer energy to an adjacent body with a lower intensity of movement. This process of transfer of thermal energy is known as *conduction,* and we define a quantity known as *temperature* as a measure of the ability of a body to transfer thermal energy to adjacent bodies by the conduction process. If the temperature of a body is higher than that of adjacent bodies, heat will be conducted from it; if it is lower, the reverse is true. We conveniently choose a scale of temperature in terms of certain transitions that occur in nature. Specifically, we define the melting point of ice as zero degrees centigrade (0°C) and the boiling point of water as 100 degrees centigrade. In energy conversion processes involving thermal energy, it is convenient to define an alternative temperature scale, commonly referred to as the scale of *absolute temperature*. Here, the measure of temperature is the kelvin (K) rather than the degree centigrade. Zero kelvin corresponds to -273.17°C and is the condition in which all atomic and molecular motions have effectively ceased.

In a system that does not receive energy from or emit energy to the outside, the *total amount of energy can be increased only by converting mass into energy via nuclear processes*. In the absence of these processes, *the total amount of energy remains constant* (this is the basis of the first law of thermodynamics). However, within the given system, the *form* of energy may change (e.g., chemical energy may be converted into thermal energy or thermal energy may be converted into mechanical energy). Before discussing these conversion processes, we shall digress briefly to discuss and explain the units by which energy is measured, since these are vital in what follows in this book.

1.1.2 Units of Energy

In this book we shall use the now widely accepted Système Internationale (SI) units of energy. Here the basic unit of energy is a *joule*. The magnitude of the joule may be understood from the following examples for various types of energy.

Kinetic energy: A mass of 2 kilograms (4.4 Ib) moving at a velocity of 1 meter per second (3.3 ft/s) has a kinetic energy of 1 joule.

Potential energy: A mass of 0.1 kilogram (3.5 ounces) at a height of 1 meter (3.3 ft) above the earth's surface has a potential energy of 1 joule.

Chemical energy: Burning 1 kilogram (2.2 Ib) of coal releases approximately 3.5 million joules of energy.

Electrical energy: A 100-watt lamp burning for 1 second uses 100 joules of electrical energy.

Nuclear energy: Converting 1 kilogram of mass into energy releases 80 thousand million million joules.

Thermal energy: Heating 1 kilogram of water by 1°C (1 .8°F) requires 4187 joules.

The rate of energy flow or production is measured in *watts*, I watt being I joule of energy per second.

Units such as the joule and the watt are rather small for many practical purposes. In the SI system of units the practice is to use prefixes to denote larger quantities. Thus:

1 kilojoule (kJ) = 1000 joules
1 megajoule (MJ) = 1 million joules
1 gigajoule (GJ) = 1 thousand million joules
1 terajoule (TJ) = 1 million million joules
1 kilowatt (kW) = 1000 watts
1 gigawatt GW) = 1 thousand million watts

Many other measures of energy are in common use, and it may be helpful to state here the relationship between these units and their SI equivalents:

1 calorie (energy required to heat 1 gram of water by 1°C)	= 4.187 joules
1 British thermal unit (Btu) (energy required to heat 1 lb of water by 1°F)	= 1055 joules
1 therm (100,000 Btus)	= 105.5 megajoules
1 mtce (energy released by burning 1 million tons of coal)	= 26,892 terajoules

1.1.3   Energy Conversion Process

The extent to which one form of energy can be converted into another is limited by practical considerations. The fraction converted in a given process is often referred to as the *efficiency* of the process. Thus, in converting x units of energy in form A to y units in form B, the percentage efficiency is defined as $100y/x$. The energy not converted to form B (i.e., $x-y$ units) may remain in form A or may find its way into other forms (C, D, etc.) as a result of the process.

An example of energy conversion leading to power generation is hydroelectric power generation. The *potential energy* of the water in a mountain reservoir or lake is first converted into *kinetic energy* of a turbine, which in turn is converted into *electrical energy* by means of a generator. All of these energy conversion processes are quite efficient; with good design they might even approach 100% efficiency. The energy not converted to electrical energy in this process is mainly dissipated by increasing the thermal energy of the water leaving the power station.

Another common example of energy conversion is that of converting the *chemical energy* of fossil fuels (e.g., coal or oil) into *electrical energy* through the medium of a conventional power station. This case is illustrated in Figure 1.2. Suppose that we start with 100 GJ* of chemical energy in the form of coal. This energy is released at a high temperature (typically 2000°C). Some of the energy (typically 10 GJ) leaves the power station as thermal energy in the flue gases going up the stack. However, most is transferred by thermal radiation and con-

*One gigajoule of energy would be sufficient to power a 100-watt light bulb for 116 days (nearly 4 months).

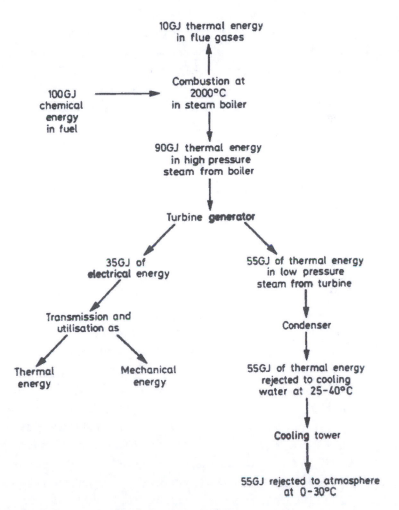

Figure 1.2: Energy conversion in a power station.

vection to the water in the boiler tubes, converting this water into high-pressure steam at perhaps 500°C. This steam is passed into a turbine, where some of the *thermal energy* is converted into *electrical energy* (typically 35 GJ), and the rest of the original energy (55 GJ) is rejected as thermal energy into lukewarm cooling water at 25–40°C.

Thus, only about one-third of the original chemical energy in the coal has actually been converted into a useful alternative form, namely, electrical energy. The efficiency achievable in the conversion of thermal energy (the intermediate form of energy in the process described here) into electrical energy is determined by the temperature range over which the process operates. If we were able to reject the heat at a temperature close to absolute zero, the residual thermal energy would be negligible. However, we are forced to reject the energy at a temperature slightly above that of our normal surroundings, at which temperature a very large amount of the thermal energy still remains. Thus, if we lived on a planet where the ambient conditions were close to absolute zero, our energy conversion efficiencies could be made much higher, though there would be some other difficulties. This basic limitation on the conversion of thermal energy into other useful forms is fundamental to thermodynamics.

We may make better use of the chemical energy used in power generation if we can use the thermal energy leaving the station directly, for example, for domestic or industrial heating. However, this heat is not very useful at the lukewarm temperatures of the cooling water. A *combined heat and power* (CHP) station rejects heat at a higher and more useful temperature (100°C, say), but, for the reasons explained above, this leads to some reduction in the electrical output of the station. This trade-off between heat and power generation can sometimes be economic, particularly where there is a large demand for heat. Thus, CHP stations have found extensive application for power generation and district heating in the Scandinavian countries and Russia.

A device for converting thermal energy into another form of energy (kinetic, potential, electrical, etc.) is referred to as a *heat engine*. A typical heat engine would be the turbine of a power station. Other examples are the jet engine of an airplane and the internal combustion engine of an automobile. All these devices take thermal energy generated at a temperature T_1, carry out some form of energy conversion, and reject the residual thermal energy at a lower temperature, T_2. Here, T_1, and T_2 designate temperatures on the absolute (kelvin) scale of temperature. The *maximum* efficiency η obtainable from any heat engine is given by an equation first derived by Carnot in 1824:

$$\eta = \frac{T_1 - T_2}{T_1}$$

This equation shows that if T_2 is zero on the absolute scale, the efficiency can theoretically approach unity (i.e., 100%). However, in practice it is necessary to reject the heat at a temperature somewhat above normal ambient temperature (e.g., 300 K or 27°C). Thus, the maximum efficiency is likely to be around 50–60% in a common heat engine, with practical efficiencies being lower than this because of departures from ideal behavior.

1.2 EARTH'S INTERNAL HEAT GENERATION

The classical view of the origin of the planet Earth was that it was formed from material torn out of the Sun, possibly by the gravitational pull of a star that passed close by. The material torn out would be initially in gaseous form and would then condense into a liquid, which would solidify on its outer surface, forming Earth's crust. This view is now considered to be unlikely because the materials of which Earth is made (iron, calcium, magnesium, aluminum, etc.) do not normally occur inside stars like the Sun. Planets like Earth are, in fact, something of an oddity in the whole collection of galactic material, which consists mainly of elements such as hydrogen and helium. To create heavier elements such as carbon and neon by fusion of lighter elements, a temperature of 200 million °C is required, and even heavier elements (iron, cobalt, nickel, etc). require temperatures of 4500 million °C for their formation. Such temperatures do not exist in the Sun, but they have been postulated to occur in "supernovaes," explosions of great violence in which giant stars end their lives.

Hawking (1988) has described the process of creation of our universe (Figure 1.3). When the "big bang" occurred, the universe was infinitely small and also infinitely hot. Seconds later the temperature had fallen to 10^{10} degrees and the initial expanding universe consisted mainly of particles; photons, electrons, protons, and neutrons. After a hundred or so seconds these particles started to combine to form the nuclei of helium, hydrogen, and "heavy" hydrogen (deuterium). This process was completed within just a few hours, and the production of helium and hydrogen then ceased. For the next million or so years the universe expanded as the temperature dropped to a few thousand degrees. Inhomogeneities developed and some of the denser regions of the gas cloud stopped expanding and started to collapse under gravitational attraction.

The collapse caused rotation, and disklike galaxies were born. Slowly regions of higher density were formed heated by fusion reactions converting hydrogen to helium, and "stable" stars were created. However, the larger the star the more rapid the consumption of hydrogen. Heavier elements like carbon and oxygen were formed as a result of the very high temperatures, but finally a crisis occurred with the exhaustion of the hydrogen fuel: the star collapsed in an explosive way (a *supernova*), and in the final stages of the supernovae explosions the heavier elements were formed. These first-generation stars lasted a few hundred million years, and the debris from their destruction together with the original gaseous cloud formed the source material for second- or third-generation stars like our Sun. These were formed about 5 thousand million years ago.

Modern theories suggest that the formation of the planets (for instance, Earth) was a multistage process. First, the solar nebula was formed by the collapse of the dense rotating interstellar gas cloud containing the debris from ear-

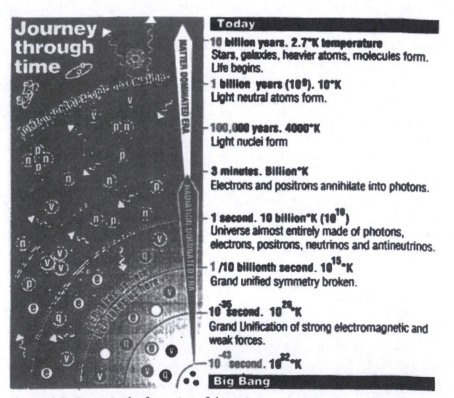

Figure 1.3: Stages in the formation of the universe.

lier supernovae condensed to form dust. In the 1950s and 1960s it was assumed that the second stage consisted of the accretion of this dust into an initially solid and cold Earth —in contrast to the classical theory in which Earth was initially a molten body that gradually cooled. It was necessary to postulate a mechanism by which the core of Earth became molten, and the explanation for this was believed to be in the behavior of the radioactive materials uranium, thorium, and potassium within Earth's core. Once temperatures were high enough, melting, followed by segregation of the molten core, was assumed to occur. However, analysis of the crustal rocks has provided an estimate of the time at which this segregation must have happened. This suggests that core formation occurred relatively early and that Earth must have accreted in a relatively hot condition.

This important new finding has led to alternative theories for the stages subsequent to the formation of the solar nebula. These more recent theories assume the dust (associated with the residual gases) accreted rapidly—within a few million years—to form larger kilometer-sized bodies, referred to as "planetesimals." These were initially hot, heated by the compression during the initial collapse phase and perhaps also by radioactive decay of short-lived isotopes like ^{26}Al. These planetesimals in turn slowly accreted into planets by mutual interaction. As the planets approached their final masses, they were impacted by larger and larger objects. Indeed it is now believed that one or more very large Mars-like objects impacted within Earth at a late stage in the accretion process. The vaporized material ejected then coalesced in orbit around Earth to form the Moon. This hypothesis can explain the absence of a metallic core in the Moon and the high angular momentum of the Earth-Moon system. This massive impact would also have completely melted Earth with surface temperatures as high as 16,000 K. It would also have completely changed the structure and composition of Earth's crust compared with that which had existed during the early accretion period.

Many atomic species were formed during the cosmic processes described above, some of which are unstable. We shall now focus on the atom, and on the particular case of uranium, which is central to our story.

Figure 1.1 shows a typical impression of an atom. It consists of a nucleus made up of dense particles called *nucleons.* These are of two main types, namely, *protons,* which each carry one unit of positive electric charge, and *neutrons,* which have the same mass as protons but are electrically neutral. Thus, the nucleus has a total electrical charge equal to the number of protons within it. Orbiting around the nucleus there is a cloud of *electrons,* which can be thought of as very small particles (compared to the nucleons). Each electron carries one unit of negative charge, and, to maintain a balance of electrical charges for the atom, the number of electrons equals the number of protons. The number of protons determines the *atomic number* of the particular species of atom of a given chemical element. The total number of nucleons (neutrons plus protons) determines the *atomic mass*. An atom of hydrogen has a nucleus consisting of only one proton and a single electron orbiting around it. The carbon-12 atom, shown in Figure 1.1, has 6 protons and 6 neutrons in its nucleus; thus its atomic mass is 12. There are 6 electrons orbiting around the carbon-12 nucleus. At the other end of the atomic mass scale, the most common form of uranium atom, uranium-238, has 92 protons plus 146 neutrons in its nucleus and 92 electrons orbiting around the nucleus.

Most of any given atom consists of space. If a hydrogen atom was magnified until it was 100 meters across, the electron would resemble a pinhead revolving around a ball bearing 50 meters away. The actual density of the material in the nucleus (the ball bearing) is incredibly high, typically 240 million metric tons per cubic centimeter.

Stable nuclei of low atomic mass tend to have about the same number of protons and neutrons, and those of higher atomic mass have about four and a half times as many neutrons as protons. Nuclei in which the ratio of neutrons to protons departs from this value are unstable, and they may undergo a spontaneous change in the direction of stability. During this spontaneous change, various forms of radiation are emitted from the nucleus:

Alpha (α) radiation is the emission of a particle having a mass four times that of the hydrogen nucleus and consisting of two protons and two neutrons. An α-particle is identical to the nucleus of the element helium.

Beta (β) *radiation* consists of very small charged particles, namely, electrons or positrons (positive electrons).

Gamma (γ) *radiation* is electromagnetic radiation that is similar in nature to light or radio waves, except that it has a very short wavelength and is capable of penetrating through a large thickness of matter.

Neutron radiation is the emission of neutrons. This occurs in a number of decay processes and can help to start the nuclear fission reaction, as described below.

The radiation arising from nuclear decay is emitted at very high velocities, typically 8000 km (or 5000 miles) per second for α, β, and neutron radiation and the speed of light for γ radiation. The creation of this kinetic energy results in a small decrease in the total mass of the system, as described above. The emitted particles collide with surrounding atoms, causing them to move and vibrate—in other words, increasing their thermal energy. Thus, the decay process leads to the generation of thermal energy.

For each *chemical species*, corresponding to a given *atomic number*, there are often several possible configurations of the nucleus characterized by different numbers of neutrons, the number of protons remaining constant. That is, several different values of the atomic mass are possible for a given *atomic number*. Each value of the atomic mass characterizes an *isotope* of the element in question. Individual isotopes are described symbolically by giving the atomic mass as a superscript before the symbol for the element; thus the most common isotope of carbon is described as ^{12}C, though small amounts of ^{13}C and ^{14}C exist in all natural forms of carbon. Similarly, the element uranium exists naturally in three isotopic forms, namely, ^{234}U, ^{235}U, and ^{238}U.

A number of the isotopes that exist in nature are unstable and are subject to decay by the process described above. When an isotope decays to another form through the emission of alpha, beta, or gamma rays, the new form may itself be unstable and may also decay. Eventually, a stable form will be reached, but many stages may have to be gone through before this is achieved. The resulting *decay chains* can be very long for the elements with atomic mass numbers above 200. The decay chain for uranium (^{235}U) is shown in Figure 1.4.

The decay processes illustrated in Figure 1.4 result in the emission of heat as the emitted rays are absorbed. These uranium isotopes were present in the material from which Earth was formed, as were a number of other unstable radioactive isotopes. An example is potassium-40 (^{40}K), which decays very slowly to argon-40 (^{40}Ar) by emitting beta radiation. Other unstable isotopes that may have been present are aluminum-26 (^{26}Al) and palladium-107 (^{107}Pd). All of these radioactive decay processes led to the release of thermal energy into Earth's material. An important parameter governing the rate of release of thermal energy in this way is the *half-life* of an unstable isotope, which is the time required for half the unstable atoms originally present to decay to their new isotopic form. Thus, after a period corresponding to 1 half-life, half the original atoms remain; after 2 half-lives, a quarter remain; after 3 half-lives, an eighth remain; and so on. After 10 half-lives, only 0.1% of the original material remains. The half-lives of ^{238}U and ^{235}U are 4500 million and 700 million years, respectively. The half-life of ^{40}K is 1300 million years. Other isotopes that were originally quite abundant on Earth include ^{26}Al, which has a half-life of 0.7

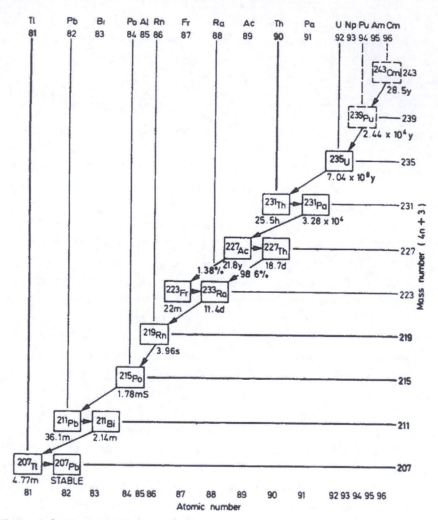

Figure 1.4: Uranium-235 decay chain.

million years, and [107]Pd, which has a half-life of 6 million years.

Although the decay of these naturally occurring isotopes is extremely slow, a very large amount of decay has occurred within the lifetime of Earth (4500 million years—comparable to the half-life of [238]U). Heat released by radioactive decay can escape from Earth only by being conducted to the surface and being radiated away into space. However, heat loss from the interior of Earth to the surface is quite small (currently about 30,000 GJ per second).

1.3 THE EARTH'S ENERGY FLOW

The heat content of the earth represents an energy store of enormous magnitude. If the average temperature of the earth was reduced by 0.001°C, energy equivalent to that available from 130 million million tons of coal would be released. This is roughly equivalent to 200,000 times the amount of coal mined in the United States in 1 year. Also for comparison, the total available resources of energy (from both fossil fuel and fission sources) are estimated to be equivalent to about 3 million million tons of coal (Armstead, 1978). Tapping this *geothermal* source of energy might be one way out of our energy difficulties, but great technological problems and costs are involved.

The temperature of the earth's core is estimated to be around 4000°C (Gass, 1971), but because of the insulating properties of the earth's solid crust, heat leaks from the center of the earth to the outside very slowly—at a rate estimated by Gass (1971) to be about 0.06 joule per second per square meter of the earth's surface. This gives a total heat out-leakage of 800 million million million (8×10^{20}) joules per year, which is equivalent to approximately 20,000 million tons of coal per year and corresponds to a very small fraction of the total heat content of the earth.

There has been a tendency for the radioactive (heat-generating) material to be concentrated in solid form in the earth's crust, contributing to the net out-leakage of heat of 8×10^{20} J/yr. The natural outflow of heat from the earth corresponds to about 30,000 GJ/s (or gigawatts), which may be compared with the total worldwide electricity consumption of about 570 GW. Both of these figures pale in comparison with the amount of energy received from the sun, which is about 170 *million* GW. The heat flows to and from the earth are shown in Figure 1.5 (Doff, 1978).

We thus see that the amounts of energy received from the sun and arising from the earth's core grossly dominate in magnitude the energy required to sustain human activities at any conceivable standard of living. The problem is not one of a *shortage* of energy but rather one of the *economics* of utilizing the energy from their sources. Both geothermal and solar sources are highly dispersed, and the capital costs involved in concentrating the sources and tapping them are enormous. Regarding the capital costs as a direct measure of the amount of human effort required to exploit energy resources, we see that we may not be able to meet our ambitions for human development with energy from these dispersed sources. When we can make use of natural concentrating mechanisms such as geothermal hot springs (for geothermal energy) or hydro-

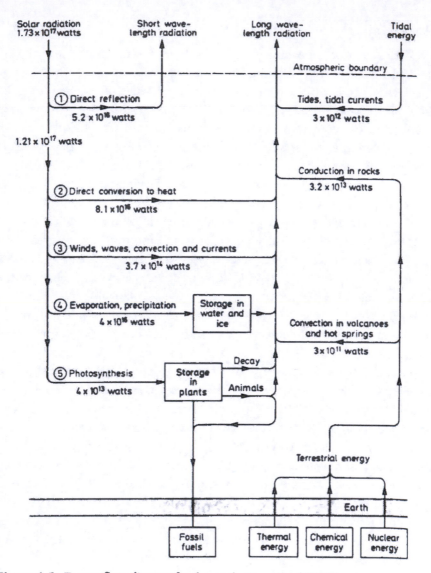

Figure 1.5: Energy flow diagram for the earth. From Dorf (1978).

electric power sources (for solar energy), we should clearly do so. However, the opportunities are limited and would not allow us to progress at the rate we desire. Specific problems with solar energy are its diurnal variations and the severe effect of cloud interference, as illustrated in Figure 1.6 (Duffie and Beckman, 1980).

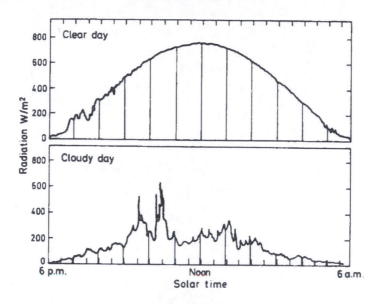

Figure 1.6: Total solar radiation to a horizontal surface for clear and cloudy days at latitude 43° for days near the equinox. From Duffie and Beckman (1980).

The optimum solution, in our view, is to use the best features of all available energy sources, and we believe that any sensible energy scenario for the earth must of necessity include one or another form of nuclear energy.

1.4   THE FISSION PROCESS

The release of energy from naturally occurring radioactive isotopes is far too slow to make them a practical energy source in themselves. However, a much more rapid release of energy is possible through the process of *nuclear fission*, which is illustrated in Figure 1.7. A neutron from a decay process may collide with a heavy nucleus (e.g., uranium), causing it to split into small nuclei (fission products) while releasing several more neutrons. These neutrons, in turn, can cause further uranium atoms to split. For a small piece of uranium this process will not be self-sustaining, because the neutrons escape from the surface. However, the bigger the piece of uranium, the greater the chance of the neutrons being absorbed, and a self-sustaining sequence (called a *chain reaction*) can be set up if a large enough mass (i.e., a critical mass) is available.

The release of energy in the fission process may be illustrated by considering

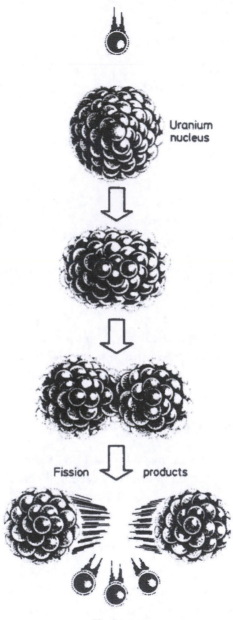

Figure 1.7: Diagrammatic view of fission process.

the fission of a ^{235}U atom, which splits up into barium and krypton atoms and releases three more neutrons:

$$^{235}U + n \longrightarrow {}^{141}Ba + {}^{92}Kr + 3n$$

If we could weigh components in this reaction we would find that those on the right-hand side of the equation weighed 0.091% less than those on the left-hand side. Thus, during the reaction, approximately 0.1% of the original mass is converted into energy. This energy appears as kinetic energy of the fission products and neutrons, which then collide with surrounding atoms and increase their thermal vibration, that is, release heat. For each kilogram of ^{238}U totally fissioned by the above reaction, 80 million million (8×10^{13}) joules are released. This is equivalent to the energy available from 3000 tons of coal.

Uranium-235 is described as a *fissile* isotope; unfortunately, naturally occurring uranium consists mainly (99.3%) of a *nonfissile* isotope, ^{238}U. Thus, only a small part of natural uranium can be burned in the fission process to produce energy. The proportion of ^{235}U in natural uranium is 0.71% by weight, and thus 1 kg of natural uranium is equivalent in energy potential to about 20 tons of coal. However, the energy potential of uranium can be increased about 100-fold by conversion of the nonfissile ^{238}U into another fissile material, namely, plutonium-239 (^{239}Pu). We shall return to this below.

The three neutrons emitted in the above fission reaction have an initial velocity of typically 20,000 km/s (about 6% of the velocity of light). Although these fast neutrons can interact with other atoms of ^{235}U, their chance of doing so can be increased by about 1000-fold if their velocity can be reduced, say, to 2km/s. These slower-moving neutrons would have a velocity similar to that of atoms vibrating due to thermal motions, and hence they are often called *thermal neutrons*. Nuclear reactors using the fast neutrons are often termed *fast reactors*, and those using the slower neutrons are termed *thermal reactors*.

Fast neutrons are converted into thermal neutrons as a result of a series of collisions with surrounding atoms. If a fast neutron hits a large atom, it tends to bounce off and lose only a small amount of its energy. However, if it hits a small atom such as hydrogen or carbon, it will lose a significant fraction of its kinetic energy. (An analogy may be made to the motion of balls on a billiard table. If a ball hits the massive cushion of the table, it bounces off with very little loss of velocity, or kinetic energy. If it hits a stationary ball, it may lose a large proportion of its kinetic energy, which is transferred to the other ball in the collision.) Thus, to convert a fast neutron to a slow or thermal neutron requires about 2000 successive collisions with uranium atoms but only about 20 collisions with the lightest atom, hydrogen. In the collision process the neutrons are sometimes absorbed without leading to a subsequent fission. Moreover, each successive collision may lead either to a fission reaction or to the neutron's combining with the atom with which it is colliding to make another isotope. Thus, there is an advantage in surrounding the uranium with lighter material that can lead to the conversion of fast neutrons to thermal neutrons, which can then pass back into the uranium; this process is known as *moderation*, and the light material used is termed a *moderator*. Moderators used in therrmal reactors have included hydrogen (in the form of its oxide, water), the hydrogen isotope deuterium (also in the form of its oxide, heavy water), and carbon (usually in the form of graphite). The best moderator is heavy water, which absorbs neutrons only weakly. However, heavy water is an expensive material and it is often preferable to use ordinary (light) water even though it absorbs neutrons much more strongly.

Because ^{238}U absorbs neutrons, it is not possible to produce a self-sustaining chain reaction by simply assembling a large enough mass of natural uranium, which is 99.3% ^{238}U. However, if pieces of natural uranium are distributed within heavy water or graphite, the neutrons produced in the fission reaction are converted to thermal neutrons (which, as mentioned above, are 1000 times more effective than fast neutrons in continuing the chain reaction), and a self-sustaining chain reaction is possible. This idea was first demonstrated by Enrico Fermi at Stagg Field, Chicago, on December 2, 1942; Fermi employed pieces of uranium distributed in a "pile" of graphite. Light water cannot be used to sustain a chain reaction with natural uranium because of the high absorption of neutrons by hydrogen. However, nuclear reactors may be constructed with light water as a moderator provided the concentration of ^{235}U is increased from 0.71 to about 3%.

As we shall see later, various generic types of nuclear reactors have arisen from the various possible combinations of fuel and moderator. These can be classified as follows:

1. Heavy water–moderated, heavy water–cooled reactors. These are the basis of the Canadian line of development and are called CANDU reactors.
2. Graphite-moderated, gas-cooled natural uranium reactors. These are the basis of the British Magnox reactors.
3. Light water–moderated, light water–cooled reactors with fuel enriched in uranium-235. These are the basis of the U.S. boiling-water reactor (BWR) and pressurized-water reactor (PWR) development.

The further development of the British system, the advanced gas-cooled reactor (AGR), uses graphite as a moderator and a somewhat enriched fuel to compensate for the fact that the fuel is contained in stainless steel, which absorbs a significant fraction of the neutrons.

We shall describe and discuss all the above reactors, and in particular their cooling problems, in the following chapters. However, before doing so, it may be interesting to glance back to prehistoric times. The light water–cooled and –moderated reactors have been around far longer than one might imagine. In fact, the invention of Enrico Fermi was preempted by nature approximately 2 billion years earlier. In 1972, evidence was found of the dormant remains of a natural fission reactor located at Oklo, in the West African Republic of Gabon. This natural reactor operated for a period of hundreds of thousands of years. Its existence was discovered by an intriguing piece of detective work by French nuclear scientists.

In May 1972, H. Bouzigues obtained a curious result during a routine analysis of standard samples of uranium ore from Gabon. He found that they contained about 0.4% less ^{235}U than expected. This was not due to an error in his analysis or to a natural variation. On this planet, at any particular time, the ratio of ^{235}U to ^{238}U is fixed; some other explanation had to be found for the discrepancy. A careful investigation carried out by the French Commissariat à l'Energie Atomique (CEA) traced the abnormal ore to one particular location in Oklo. It was concluded that the deficiency in ^{235}U could be explained only by the occurrence of a natural fission reaction at the site. At the time this natural reactor was operating, the ore was buried deep underground and natural groundwater served as a moderator and to some extent as a coolant. Such a reactor would not be possible with the present-day concentration of ^{235}U in naturally occurring uranium, as we explained above. However, it should be remembered that the half-life of ^{235}U is about 700 million years and that of ^{238}U is about 4.5 billion years. Thus, in prehistoric times, the concentration of ^{235}U was much higher than it is today. When the earth was formed some 4.6 billion years ago, the concentration of ^{235}U in natural uranium was about 25%, and it had decreased to about 3% at the time when the Oklo reactor was operating.

It is thought that the natural reactor at Oklo operated under considerable pressure and temperature and that the rate of reaction was controlled by variations of the water (moderator) density. Cooling was provided mainly by conduction, with some limited circulation by permeation. The power level is estimated to have been somewhat less than 100 kW and the total energy released over the period of operation to have been about 4.7×10^{17} joules (15,000 MW-years), representing the fission of about 6 metric tons of ^{235}U.

This amount of energy is about that released in a modern pressurized-water reactor in 4 years.

It is possible that a combination of local circumstances may have led to other naturally occurring reactors. Though the search continues, none has been located so far. Such naturally occurring reactors have been impossible for the past 2 billion years because the ^{235}U concentration has been below the required 3%. A detailed review of the Oklo reactor phenomenon is given by Cowan (1976). It is interesting to note that 2 tons of plutonium-239 would have been produced at the Oklo natural reactor, though the amount that remains is infinitesimal because of the comparatively short half-life (25,000 years) of ^{239}Pu. Thus, it cannot be claimed that plutonium is an entirely human creation.

1.5   THERMAL ENERGY RESOURCES

To put energy production from uranium into context, it is interesting to compare the known recoverable resources and current use of uranium with the recoverable resources and current use of fossil fuels (coal, oil, and natural gas). Figures provided from the 1992 World Energy Congress in Madrid are shown in Tables 1.1 to 1.3.

The estimated total fossil fuel resources (Table 1.2) are 209 x 10^{21} joules, of which 40.3 x 10^{21} joules are estimated to be recoverable.

Table 1.1 • **Proven Fossil Fuel Reserves and Production Rates**

Fuel	ESTIMATED PROVEN RESERVES IN Gtoe*	10^{21}J	Estimated Production to Date in Gtoe*	Life at Current Production Rate in Years
Coal (excluding lignite)	496	23.5	n/a	197
Lignite	110	5.2	n/a	293
Oil	137	6.5	86	40
Natural gas	108	5.1	40	56

Source: WEC 1992 *Survey of Energy Resources.*

*1 Gtoe = energy equivalent to 10^{12} tons of oil.

Table 1.2 • Ultimately Recoverable Fossil Fuel Resources

Fuel	ULTIMATE RESOURCES IN		
	Gtoe	10^{21}J	Percent
Coal and lignite	3400	161.6	76
Conventional oil	200	9.5	5
Unconventional oil			
Heavy crude	75	3.6	2
Natural bitumen	70	3.3	2
Oil shale	450	21.4	10
Natural gas	220	10.4	5
Total (approx.)	4400	209.1	100

Sources: WEC 1992 *Survey of Energy Resources;* WEC 1989 *World Energy Horizons 2000–2020;* C.D. Masters et al., *World Resources of Crude Oil and Natural Gas,* 13th World Petroleum Congress, 1991; C.D. Masters et al., *World Reserves of Crude Oil, Natural Bitumen and Shale Oils,* 12th World Petroleum Congress, 1987.

Table 1.3 • Uranium Resources

Source	Amount, M Tons	Energy content in Thermal Reactors, 10^{21}J	Life at Current Production, Years	Energy Content in Fast Reactor, 10^{21}J
"Known" uranium resources	3.12	1.37	76	68.5
Speculative resources	8–13	3.5–5.7	—	175–285
Extraction from seawater	~4000	1760	—	88,000

Sources: OECD/IAEA "Red Book," 1991; UKAEA, 1976.

Uranium resources can be similarly estimated, and the results are presented in Table 1.3.

The estimated "known" uranium resources are equivalent to 1.4×10^{21} joules if used in thermal reactors and 70×10^{21} joules if used in fast reactors (the difference between these values is explained in Chapter 2). Taking account of speculative resources, the energy content for utilization in fast reactors is estimated to be between 175 and 285 x 10^{21} joules. Uranium resources would be greatly increased by extracting uranium from seawater, in which it occurs at a concentration of 3 parts per billion. This extraction is not likely to be economically feasible in the foreseeable future, however.

Because "known" reserves will last only 76 years at present usage rates, it is important to utilize the uranium resources more efficiently than by simply burning them in thermal reactors. Efficient utilization of these resources can be obtained by using fast reactors. An alternative approach might be to use thermal reactors in which the excess neutrons are employed to convert thorium (placed in the reactor core) to an alternative fissile material, uranium-233. While such reactors can be made self-sustaining (i.e., they can produce as much fissile material as they consume), they cannot be designed to breed efficiently, i.e., produce enough excess fissile material to start a new reactor within a reasonable time.

Primary energy usage from 1960 to 1990 and estimated up to 2100 is given in Table 1.4 in Gtoe/year. It will be seen that nuclear power currently provides just

Table 1.4 • **Primary Energy Usage (in gigatonne of oil equivalent per annum)**

Fuel	1960	1990	2020 (est.)	2050 (est.)	2100 (est.)
Coal	1.4	2.3	2–5 ⎫		
Oil	1.0	2.8	3–4.6 ⎬	8.7–15.6	3–17
Natural gas	0.4	1.7	2.5–3.6 ⎭		
Nuclear	—	0.4	0.7–1	1.2–3.8	2.2–12
Large hydro	0.15	0.5	0.7–1 ⎫		
Traditional	0.5	0.9	1.1–1.3 ⎬	3–4	8–10
"New" renewables	—	0.2	0.6–1.3 ⎭		
Total (Gtoe)	3.3	8.8	11–17	15–27	20–42

over 5% of world primary energy demand and this is predicted to grow to between 8% and 30% by 2100.

REFERENCES

Armstead, H.C.H. (1978). *Geothermal Energy*, Chapman & Hall, London.
Cowan, G.A. (1976). "A Natural Fission Reactor." *Sci. Am.*, July, 36–47.
Dorf, R.C. (1978). *Energy Resources and Policy*. Addison-Wesley, Reading, Mass.

Duffie, J.A., and W.A. Beckman (1980). *Solar Engineering of Thermal Processes.* Wiley, New York.

Gass, I.G. (1971). *Understanding the Earth.* Artemis Press.

Hawking, S. (1988). *A Brief History of Time,* Bantam, New York.

NEA/IAEA (1991). *Uranium, Resources, Production and Demand.* Joint report by OECD and IAEA.

Newson, H.E., and J.H. Jones (1990). *Origin of the Earth.* OUP/Lunar and Planetary Institute, Houston.

U.K. Atomic Energy Authority (1976). *Uranium from Seawater.*

Wetherill, G.W. (1990). "Formation of the Earth." *Ann. Rev. Earth Planet. Sci.* 18:205–56.

———— (1980). "Formation of the Terrestrial Planets" *Ann. Rev. Astron. Astrophys.* 18:77-113.

World Energy Conference (1992). *Survey of Energy Resources.*

EXAMPLES AND PROBLEMS

1 Chemical energy release

Example: Liquid within a chemical reaction vessel releases chemical energy at the rate of 1.5 kW. The stirrer requires a further power input of 1 kW. If the vessel is cooled by water flowing in a jacket surrounding the vessel, what must the water flow rate be to limit the temperature rise across the jacket to 10°C?

Solution: The energy gained by the cooling water must equal that liberated by the chemical reaction and the heat put in by the stirrer. Let W be the water flow rate in kilograms per second. The specific heat of water is 4187 kJ/kg°C. Therefore,

$$W \times 10 \times 4187 = 1500 + 1000 \, J\,/\,s$$
$$W = 0.06 \, kg\,/\,s$$

Problem: For the chemical reaction described in the above example, calculate the rise in cooling-water temperature that will occur if the flow rate of water in the cooling jacket is set at 0.1 kg/s.

2 Pumped storage scheme

Example: A massive pumped storage scheme in North Wales has an upper reservoir containing 6.7 million m^3 of water 500 m above the lower reservoir. The scheme can supply 1800 MW(e) to the electricity grid. Assuming there are no losses, how long can the plant remain in operation at maximum output?

Solution: The total mass of water in the upper reservoir is the volume of water (6.7×10^6 m^3) multiplied by the density of water ($100 \, kg/m^3$), or 6.7×10^9 kg. The total potential energy is this mass times the head (in meters) times the acceleration due to gravity, or

Total potential energy $= 6.7 \times 10^9 \times 500 \times 9.81$

$$= 32.8 \times 10^{12} \text{ J}$$

The power output is 1800 MW(e) $= 1.8 \times 10^9$

$$= 1.8 \times 10^9 \text{ J/s}$$

Therefore, the plant can remain in operation for

$$\frac{32.8 \times 10^{12}}{1.8 \times 10^9} = 18.2 \times 10^3 \text{s}$$

$$\cong 5 \text{h}$$

Problem: Suppose that the scheme described in the example is to be extended to supply 2000 MW(e) for a period of 7 h. Assuming a conversion efficiency of 95%, calculate the volume of the upper reservoir now required.

3 National fast reactor policy

Example: It is sometimes stated that the introduction of the fast reactor allows a country to be independent of uranium imports. Why does this statement need considerable qualification? A country has a nuclear power program based on thermal reactors of 10GW by the year 2020. The demand for electric power is growing at 3% per year, and it is decided to introduce fast reactors capable of breeding sufficient plutonium for a further fast reactor in 40 years (the doubling time). What are the consequences?

Solution: Let us assume that sufficient plutonium is available from the operation of the thermal reactors (for about 20 years) to allow 10 GW of fast reactors to be installed in 2020. However, the fast reactor capacity can then increase only at the rate at which plutonium becomes available to provide fuel for further fast reactors. This rate is 1/40 per year, or 2.5%/yr. Thus the rate at which fast reactors can be installed is *lower* than the electric power demand growth rate (3%/yr). The difference must (presumably) be made up by installing further thermal reactors, which will *increase* the country's need for imported uranium. Even if a fast reactor with a shorter doubling time (say, 30 years) is available, a finite time will still be needed to replace thermal reactors by fast reactors. Thus it will take about *twice* the doubling time, i.e., 60 years, for the country to be free from uranium imports.

Problem: Suppose the country described in the example decided on an intensive program of energy conservation to reduce its growth in demand for electric power to 2%. How would this reduction affect the scenario for introducing fast reactors?

BIBLIOGRAPHY

Addinall, E., and H. Ellington (1982). *Nuclear Power in Perspective*. Kogan Page, London, 214 pp.

Eden, R., and M. Posner, (1981). *Energy Economics. Growth, Resources and Policies*. Cambridge University Press, Cambridge, 455 pp.

Haywood, R.W. (1975). *Analysis of Engineering Cycles*. Pergamon, Elmsford, N.Y.

International Nuclear Fuel Cycle Evaluation (INFCE) (1980). *Fuel and Heavy Water Availability*. Report of INFCE Working Party Group 1. International Atomic Energy Agency (IAEA), 312 pp.

Loftness, R.L. (1978). *Energy Handbook*. Van Nostrand Reinhold, New York.

Lyttleton, R.A. (1982). *The Earth and Its Mountains*. Wiley, New York.

Marshall, W., ed. (1983). *Nuclear Power Technology*. Clarendon, Oxford.

McMullan, J.T., and R. Morgan, (1983). *Energy Resources*, 2d ed. Edward Arnold, London,191 pp.

Merrick, D., and R. Marshall (1981). *Energy, Present and Future Options*, vol. 1. Wiley, New York, 340 pp.

Mustoe, J. (1984). *An Atlas of Renewable Energy Resources: In the United Kingdom and North America*. Wiley, New York.

2
How Reactors Work

2.1 INTRODUCTION

In Section 1.4 we briefly introduced the *fission* process and explained that it leads to the generation of heat within the nuclear fuel. This heat can be used to generate electrical energy in a nuclear power station. In this chapter we shall further explore this heat generation process and discuss the aspects of nuclear reactor design concerned with removing and utilizing the heat.

2.2 THE FISSION PROCESS

Given enough fissile material, such as ^{235}U, fission leads to the production of a self-sustaining chain reaction in which the neutrons arising from a given fission cause other fission reactions, which in turn cause others, and so on. Each fission reaction produces either two or three neutrons (with an average of about 2.5 neutrons per fission). Since only one neutron is required to cause a fission, about 1.5 neutrons are available in excess. In a *supercritical* system, these neutrons progressively increase the rate of fission, which is the basis for an atomic bomb. In a nuclear reactor the excess neutrons are either absorbed or used to produce more fissile material. Thus, a nuclear reactor has a *critical* mass of fissile material in which a state is achieved where, on average, one of the neutrons arising from a fission causes just one further fission. We thus have a delicate balance from which a slight deviation would cause the chain reaction either to die away or to accelerate. Fortunately, there are inherent features of the nuclear reaction within nuclear reactors that prevent the uncontrolled acceleration of the fission process and allow control of the reactor. We shall return to this matter when we discuss the component parts of nuclear reactors in Section 2.3.

Let us look at the chain reaction within a nuclear reactor in a little more detail. Several different processes are occurring. We can categorize them as:

- Those immediate processes related to the fission reaction itself—the so-called *prompt* processes:

—The fission of the U-235 atom with the formation of two (or more) fission products and the release of energy and two or three neutrons
—The emission of radiation in the form of β and γ radiation

- Those processes which effectively "lose" neutrons from the system:

—The absorption of neutrons in the inert U-238 to form a new element, Pu-239, which in turn can be fissioned by an incident neutron
—The absorption of neutrons in fission products or structural materials: we call this *parasitic* absorption
—The leakage of neutrons outside the system; these are "lost" neutrons absorbed in the surrounding shielding

- Finally, those processes which do not occur immediately—the so-called *delayed processes*; these include:

—The emission of delayed neutrons from transmuting fission products
—The release of other radiation (β and γ radiation) from fission products

As we can see, there are various fates for fission neutrons; so what is the overall balance in an operating reactor?

Figure 2.1 illustrates the fate of fission neutrons in a thermal reactor. Here 100 fissions produce, on average, 259 neutrons. These processes, which effectively "lose" neutrons from the system in *parasitic* absorption, account for 59 neutrons. The remaining 200 neutrons undergo interactions with the fuel as illustrated. Some lead to further fissions giving a steady state value of 100 fissions with which to continue the process. The original fuel in the reactor consists of a small amount of fissile U-235 and a much larger amount of inert U-238. The U-235 absorbs 78 of the remaining 200 neutrons, producing 63 fissions; the other 15 neutrons are absorbed to produce nonfissile U-236. The inert U-238 absorbs 63 of the original neutrons, but only 5 of these result in fissions (U-238 can only be split by very high velocity neutrons). As we have seen, the process of absorption of 58 of the 259 fission neutrons in U-238 gives rise to a new element, plutonium-239. This very important process of plutonium production is actually quite complex. First U-238 is transformed into U-239, which has a 23.5-min half-life and decays by beta emission to another man-made element, nep-

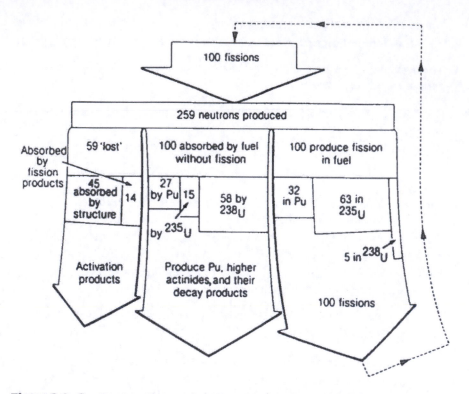

Figure 2.1: Creation, reaction, and absorption of neutrons during fission in a thermal reactor.

tunium-239 (NP-239). This in turn decays by beta emission, with a 2.3-day half-life, to Pu-239. Plutonium is comparatively stable and has a 24,000-year half-life.

In a reactor operating in the steady state, Pu-239 exists in the fuel as a result of the absorption process described above. It is a fissile isotope and interacts with the remaining 59 neutrons (out of the original 259) to produce 32 fissions, the remaining 27 neutrons being absorbed to form higher plutonium isotopes (Pu-240, Pu-241, etc.).

Thus, in the steady state, about 30% of the energy produced by a thermal nuclear reactor is actually being produced by the fission of plutonium.

The neutron events associated with a fast reactor are illustrated in Figure 2.2. Here, the fast-neutron fissions produce more neutrons than do thermal neutron fissions, namely, 292 instead of 259. Furthermore, fewer neutrons are lost in the fission products, the structure, and leakage. The 253 neutrons that are not lost interact with the fuel, which in a fast reactor consists of a mixture of approxi-

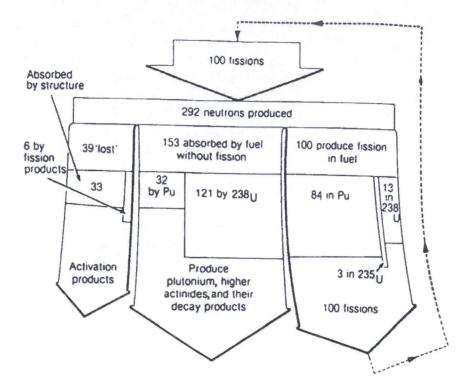

Figure 2.2: Creation, reaction, and absorption of neutrons during fission in a fast reactor.

mately 20% plutonium and 80% *depleted* uranium, i.e., natural uranium from which some of the U-235 has been extracted. The uranium is depleted because it has either been through a thermal reactor, where the U-235 has been largely burned, or passed through an enrichment plant, where the U-235 has been preferentially separated out for utilization in thermal reactors. Thus, the remaining U-235 contributes only 3 of the 100 fissions that take place. The bulk of the fissions come from plutonium, where 116 of the original 292 neutrons are absorbed, giving rise to 84 fissions. Uranium-238 absorbs 121 of the original 292 neutrons to produce Pu-239 by the process described above. The remainder of the original neutrons produce 13 fissions from U-238; this is higher than the comparable number for a thermal reactor because of the high energy of the neutrons in a fast reactor.

If we look at Figure 2.2 a little more closely, we see that 116 neutrons interact with the plutonium and 134 interact with U-238. Of the latter neutrons, 13 produce fission of the U-238, leaving 121 atoms of U-238 converted into Pu-239. Thus, Pu-239 is being created at a greater rate than it is being consumed. The reactor can therefore be regarded as *breeding* the fissile material Pu-239. There is a net gain in fissile material, even taking into account the fission of the small amount of U-235 consumed. This remarkable process, by which the very large amount of available depleted uranium is consumed, represents a vast energy resource.

In Chapter 1 we stated that each kilogram of U-235 totally fissioned in a reactor would release 80 million million (8×10^{13}) joules of thermal energy. This means that a typical nuclear power station producing 1000 MW of electric power would burn 3.5 kg of U-235 each day. That is, for each unit of electrical power produced, we have to produce 3.5 units of thermal energy, the remaining 2.5 units of energy being dissipated into the lukewarm cooling water.

Let us now consider how this 80 million million joules of thermal energy from the fission of 1 kg of U-235 is distributed. Table 2.1 lists the end products of the fission reaction and shows the breakdown of the energy released per kilogram of uranium for steady state reactor operation. The greatest part of the energy is released in the form of the kinetic energy of the fission products. When a fission takes place, the fragments fly apart and hit the other molecules of the system, increasing their thermal vibration (i.e., releasing heat). The neutrons and gamma radiation from the fission process also interact with the surrounding matter, inducing thermal vibrations; this amounts to about 10% of the fission product interaction. Both of these processes occur at the time of the fission reaction and are therefore called *prompt* processes. However, the fission products themselves may be radioactive, and their decay may release further energy; this represents a *delayed* release of the energy arising from fission. The fission products emit beta and gamma radiation, and, associated with the beta radiation, there is a simultaneous emission of tiny uncharged particles called *neutrinos*. As shown in Table 2.1, the energy carried by the neutrinos is significant, but since these particles do not interact with matter, this energy is lost to the system. Table 2.1 gives a clue to one of the most important technical questions in the design of nuclear power plants. The delayed release of heat due to fission product decay continues after the fission reaction has been closed down. The rate of this heating falls quite rapidly after the shutdown of the fission reactor, as shown in Table 2.2. After 1 s, the power has dropped to 6.5% of the steady state value; after 1 h, down to 1.4%; after 1 year, down to 0.023%; and so on. These power levels are small compared with the full-power level, but they are quite significant in absolute terms. For example, a reactor generating 3400 MW of thermal energy will still be producing 217 MW 1 s after shutdown, 47.6 MW 1 h after shutdown, and 0.78 MW 1 year after shutdown. Thus, it is essential to continue to cool the reactor after shutdown, and even to cool the fuel when it has been removed from the reactor. Removal of decay heat is a very important consider-

Table 2.1 • Distribution of Energy from Fission of 1 kg of Uranium–235

	Energy (10^{12} J)	
Prompt process		
fission products	69	
fission neutrons	2	
γ radiation	3	
		74
Delayed process		
β radiation	3	
γ radiation	3	
		6
Other process		
neutrinos		5
Total		85

Table 2.2 • Decay Heat Rates Following Shutdown for a Pressurized Water Reactor

Cooling Time	Percent of Steady Power at Shutdown
1 s	6.5
10 s	5.1
100 s	3.2
1000 s	1.9
1 h	1.4
10 h	0.75
100 h = 4.17 days	0.33
1000 h = 1.39 months	0.11
8760 h = 1 year	0.023

ation in reactor safety analysis and will be discussed in Chapter 4. The heat released immediately by the fission reaction and in a delayed fashion from the decay of the fission products is essentially independent of the temperature of the fuel. The temperature at which the fuel operates will be such as to drive the heat into the coolant, the steady state being that in which the rate of heat release to the coolant is equal to the rate of heat generation within the fuel.

The maximum temperature at which nuclear fuel can operate is governed by the form of the fuel; for instance, the fuel rods can be made of uranium metal, which has a relatively low melting point, or uranium oxide, which has a very high melting point. However, if there is a loss of cooling effectiveness in the reactor, these maximum temperatures may be exceeded, and in this unlikely event the fuel may melt. The consequences of such an event will be discussed in Chapter 6.

2.3 BASIC COMPONENTS OF A NUCLEAR REACTOR

Figure 2.3 illustrates schematically the principal components of a typical nuclear fission reactor.

In this typical reactor the *coolant* (high-pressure CO_2 in the AGR case chosen for illustration) at high pressure is driven by the *coolant circulator* over the *fuel element*. In many reactors (including the AGR case illustrated) this consists of pellets of uranium in oxide form sealed in a can made of stainless steel. The can (or cladding) ensures retention of the fission products so they cannot enter the coolant stream. It also prevents the coolant from attacking the fuel, which would be possible with some combinations.

The fuel elements are embodied in a structure (the *reactor core*) that allows them to be surrounded by the moderator. In the AGR case, the fuel assemblies are stacked in vertical holes (channels) in the massive structure of the graphite moderator. The whole is contained in a prestressed concrete pressure vessel retaining the high-pressure carbon dioxide gas.

The coolant extracts the heat from the fuel elements. In many reactors, this heat is then used in a boiler or steam generator to convert water to steam. In the boiling-water reactor (BWR) the steam is generated directly in the reactor core. The steam is then passed through the turbine that drives the electrical generator. The very low pressure exhaust steam from the turbine is passed to a condenser where it is converted back into water and recirculated to the steam generator (or to the reactor in the case of the BWR).

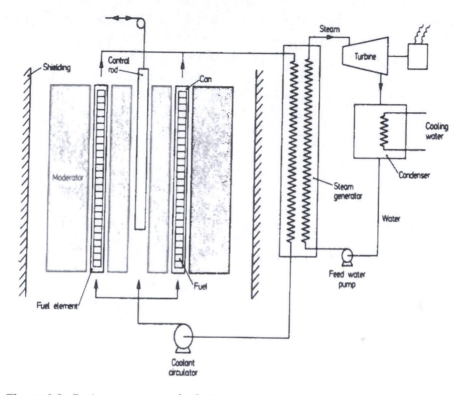

Figure 2.3: Basic components of a fission reactor.

As we saw in Chapter 1, the moderator may be a solid (e.g., graphite) or a liquid (e.g., heavy water). In light-water reactors, the coolant and moderator are both ordinary water. If the moderator is different from the coolant, it must either not react with the coolant or be separated from the coolant by a suitably intervening structure. In the heavy-water reactor, this structure is known as the *calandria*; it consists of a tank containing the heavy water penetrated by a series of tubes in which the fuel is mounted and through which the coolant passes.

The remaining main feature of the nuclear reactor core is the means of controlling neutron population, namely, the *control rods*. These consist of neutron-absorbing material such as boron or cadmium.

The number of neutrons produced per neutron absorbed is often referred to as the *multiplication factor k*. If k is the greater than unity, the neutron population increases; if k is exactly unity, the neutron population remains the same; and if k is less than unity, the population decreases. The rate of growth of the neutron population depends on the *neutron lifetime*, i.e., that time between the creation of a neutron and its interaction with the fissile material to create further neutrons.

Most of the neutrons present in the reactor are the so-called prompt neutrons. In thermal reactors they have a lifetime of typically 0.0001 to 0.001 s; in fast reactors their lifetime is even shorter. If the neutron population consisted of only these neutrons, it would grow very rapidly as soon as k slightly exceeded unity, and the reactor would be very difficult to control. This is because the time between successive generations is very short, and very rapid multiplication of the neutrons would be inevitable. For instance, for a neutron lifetime of only 0.005 s, the neutron population would increase (for $k = 1.005$) by over 20 times in 1/3 s, and this growth clearly could not be controlled easily.

Fortunately, at the steady state not all of the neutrons are of the prompt type; a small fraction (~0.7%) are of the *delayed* type, whose lifetime (as defined above) is typically 0.6 to 80 s. These delayed neutrons arise from the decay of fission products rather than directly from the fission process itself. Thus, at steady state only 99.3% of the neutrons are of the prompt type and the population is "topped up" by delayed neutrons, whose number is just sufficient to maintain the steady state, i.e., $k = 1.000$.

The control system operates essentially on these delayed neutrons, and the response of the system is such that control rod movements over a time scale of 10–20 s can give adequate control over the chain reaction.

The system is designed so that the k value cannot exceed a critical value (1.007 for the example cited above) above which the k value for the *prompt neutrons alone* is greater than unity. If k were allowed to exceed this value, rapid growth of the prompt neutron population would occur and the system would be in what is known as the *prompt critical* condition. However, the design of nuclear reactors is such that this condition is avoided.

The nuclear fission process results in intense radiation. The fission products also contribute substantially to the radiation field, and they continue to emit radiation after the fission reaction is closed down. Thus, it is very important to provide proper *shielding* around the reactor core. This shielding takes the form of a thick concrete *biological shield*. In the AGR plant illustrated in Figure 2.3 the prestressed concrete pressure vessel doubles as the biological shield.

Where necessary—as it is for water-cooled reactors—further protection is provided by housing the whole system inside a leak-tight containment building. We shall discuss the role of this containment building in possible nuclear reactor accidents in Chapters 5 and 6.

Figure 2.3 gives a generalized view of the components of one type of nuclear reactor, and it should be realized that there are many possible permutations of fuel type, coolant type, cladding, moderator, and steam generator. It would be tedious to describe every nuclear reactor type that has been built and practically impossible in any book of reasonable size to describe all those that have been conceived. Many of the early concepts for nuclear reactors departed from the format shown in Figure 2.3 in that they proposed to use the fuel in a fluid form, circulate it through the core, and pass it through heat exchangers externally before returning it to the core. The concepts included systems in which solutions of uranium salts were circulated through the core, slurries of fuel were made and circulated, or the fuel was circulated in fused-salt form or in solution in liquid metals. There was a tradition at Harwell* that in the early days it was possible to invent a reactor system in the bath in the morning and have a project by lunchtime. It took some years to realize that reactors that you have just thought of are simple, cheap, and reliable, whereas those you are actually working on are always complicated, expensive, and troublesome.

In the remainder of this chapter we shall concentrate on describing some of the main systems that have been implemented in practice and that form the basis of the development of nuclear power. These are the British Magnox and AGR (advanced gas-cooled reactors), the U.S. light-water reactors (BWR and PWR), the Canadian CANDU reactor, the Russian boiling-water graphite-moderated RBMK-type reactor, and the liquid-metal fast reactor.

2.4 THERMAL REACTORS

Although other coolants have been proposed, nearly all practical thermal power reactors are cooled with carbon dioxide (Magnox and AGR) or with light water (BWR and PWR as well as the Russian RBMK type) or heavy water (CANDU). We shall restrict the descriptions of reactors to these more common systems.

2.4.1 Natural Uranium Graphite-Moderated (Magnox) Reactors

The Magnox reactor is illustrated schematically in Figure 2.4. The coolant is carbon dioxide at a pressure of 20 bars (300 psia). The coolant is circulated through a core that consists of the moderator structure, which is built from

---

* The Atomic Energy Research Establishment of the U.K. Atomic Energy Authorities, founded in 1946.

graphite bricks containing holes through which the coolant flows and in which the fuel elements are placed. Fuel elements consist of natural uranium bars clad in cans of a magnesium alloy known by the trade name Magnox (hence the name of the reactor). The alloy does not significantly absorb neutrons, so natural uranium, rather than enriched uranium, can be used as a fuel. A typical Magnox core would be 14 m in diameter and 8 m high. The coolant gas leaves the core at 400 ºC, flows to the steam generator, and from there flows back through the gas circulator to the reactor. In the earlier designs of Magnox reactors, the pressure vessel containing the core was made of steel. In later designs it was combined with the shielding in the form of a prestressed concrete pressure vessel, which also contained the heat exchangers (in the earlier designs these were external to the pressure vessel and the shielding as shown in Figure 2.4). Magnox reactors were constructed in the United Kingdom, France, Italy, and Japan and have operated very successfully since their construction, which in some cases was around 35 years ago. The steam cycle efficiency of Magnox reactors is about

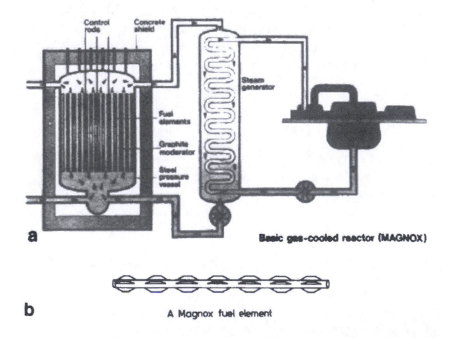

Basic gas-cooled reactor (MAGNOX)

b   A Magnox fuel element

Figure 2.4: (*a*) Carbon dioxide–cooled, graphite-moderated (Magnox) reactor using natural metallic uranium fuel; (*b*) a Magnox fuel element.

31%; this means that 69% of the nuclear heat is rejected to atmosphere via the cooling towers (Section 1.1.3).

A Magnox fuel element is shown in Figure 2.4b. The outside of the Magnox can is machined in a complex pattern of fins ("herringbone" pattern), which has been shown by detailed heat transfer experiments to be the optimal form. The swirl of the gas in the channel and the fins on the surface are an aid to heat transfer. The advantages and disadvantages of various coolants will be discussed in Chapter 3, where we shall also discuss some basic principles of heat transfer.

Although the Magnox reactor has been remarkably successful and reliable, it has disadvantages compared with some other reactor types. The principal one is its relatively low power output per unit volume of core. This leads to a large size for the core, a large investment in fuel, and high capital costs. Table 2.3 compares various reactors in terms of the average power generation rate per unit volume of the core (called the *average volumetric power density*). It also shows the rate of power generation per tonne of fuel (the *average fuel rating*) and the power generation per unit length of fuel (the *average linear fuel rating*). Compared with other reactors, the Magnox has a very low volumetric power density and a very low average fuel rating per unit mass of fuel. Both of these factors lead to high costs due to the high fuel inventory and large cores.

Table 2.3 • Volumetric Power Densities and Linear Fuel Ratings for Various Reactor Systems

Type	Reactor	Thermal Power (MW(t))	Core Diameter (m)	Core Height (m)	Core Volume (m³)	Average Vol. Power Density (MW/m³)	Average Fuel Rating (MW/tonne)	Average Linear Fuel Rating (kW/m)
Magnox	Calder Hall	225	9.45	6.40	449	0.50	—	—
	Bradwell	538	12.19	7.82	913	0.59	2.20	26.2
	Wylfa	1875	17.37	9.14	2166	0.865	3.15	33.0
AGR	Hinkley B	1500	9.1	8.3	540	2.78	11.0	16.9
	Hartlepool	1507	9.3	8.2	557	2.0	11.5	16.1
HWR	CANDU	3425	7.74	5.94	280	12.2	26.4	27.9
LWR	PWR	3800	3.6	3.81	40	95	38.8	17.5
	BWR	3800	5.01	3.81	75	51	24.6	19.0
RBMK	Chernobyl	3140	11.8	7.0	765	4.10	15.4	14.31
Fast	Phenix	563	1.39	0.85	1.38	406	149	27.0
reactor	PFR	612	1.47	0.91	1.61	380	153	27.0

2.4.2   Advanced Gas-Cooled Reactors

The low volumetric power density and low operating temperatures and pressures of the Magnox stations led to a search in the United Kingdom for an improved design. The resulting advanced gas-cooled reactor (AGR) is illustrated in Figure 2.5. In common with the Magnox reactor, the AGR uses carbon dioxide as a coolant, but the coolant pressure in the AGR is 40 bars (600 psia) and the coolant outlet temperature is 650°C. To achieve these higher temperature and pressure conditions, it was necessary to make a radical change in the design of the fuel. The fuel was changed to uranium oxide, mounted in the form of pellets inside thin-walled stainless steel tubes, which had small transverse ribs machined on the outside (Figure 2.6). These tubes (sealed at each end) were grouped in bundles of 36 (see Figure 2.6). Since the high temperatures require the use of a stainless steel can, the can material is a significant absorber of neutrons, unlike that in the Magnox reactor, and it is necessary to enrich the uranium in the fuel to about 2.3% ^{235}U (about three times the natural ^{235}U content). The AGR design benefited from the Magnox developments, particularly the design of the gas circulation system. The steam generators were mounted inside the prestressed concrete vessel, as illustrated in Figure 2.5. Since the CO_2 reactor coolant is now at a high temperature, the steam generators can be designed to provide steam under conditions similar to those found in the most efficient fossil-fuel power plant, i.e., steam at 170 bars and 560°C. This gives the AGR a considerable advantage. Its steam cycle efficiencies are around 40%, the highest of any nuclear reactor operational at present.

Referring to Table 2.3, we see that the average volumetric power density of an AGR is around three times that of the highest-rated Magnox station. The average fuel rating is also higher, by a factor of approximately 4. This leads to a more compact, capital-effective design. Nevertheless, a number of technical problems in the AGR design had to be solved. One was that the carbon dioxide coolant might react with the graphite moderator under the high temperatures and radiation fields in the reactor to produce carbon monoxide by the reaction:

$$CO_2 + C \rightarrow 2CO$$

which would corrode the graphite and reduce its strength. It was found that precise control of the carbon monoxide and water vapor content, together with the addition of methane in small concentrations, inhibited this reaction and minimized the rate of attack on the graphite. However, too high concentrations of methane and carbon monoxide could lead to carbon formation on the fuel elements, which would impair the heat transfer by reducing the turbulence

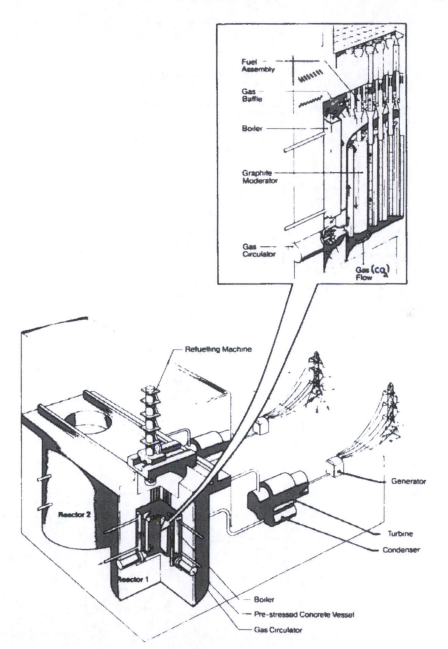

Figure 2.5: Essential features of the CO$_2$-cooled, graphite-moderated advanced gas-cooled reactor (AGR).

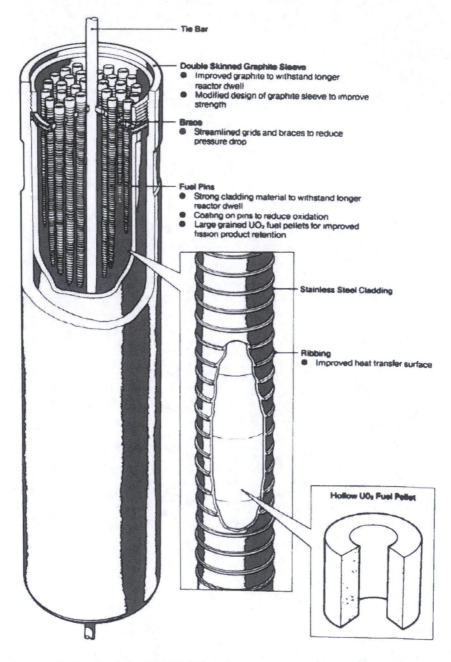

Tie Bar

Double Skinned Graphite Sleeve
- Improved graphite to withstand longer reactor dwell
- Modified design of graphite sleeve to improve strength

Brace
- Streamlined grids and braces to reduce pressure drop

Fuel Pins
- Strong cladding material to withstand longer reactor dwell
- Coating on pins to reduce oxidation
- Large grained UO_2 fuel pellets for improved fission product retention

Stainless Steel Cladding

Ribbing
- Improved heat transfer surface

Hollow UO_2 Fuel Pellet

Figure 2.6: Details of the AGR fuel element.

caused by the ribs. Fortunately, there is a range of methane and carbon monoxide concentrations (called the coolant "window") in which the satisfactory operation is possible without excessive corrosion or deposition.

An alternative design is the so-called high-temperature gas-cooled reactor (HTGR). The use of helium rather than carbon dioxide overcomes the graphite oxidation problem. Helium is inert and consequently allows higher coolant temperatures. The uranium fuel is in the form of coated particles. A kernel of low-enriched uranium carbide is coated with successive layers of pyrolytically deposited carbon and impervious silicon carbide (to retain the fission products). Two distinct lines of reactor development have been pursued. One line is the so-called pebble-bed reactor, developed in Germany, whose core consists simply of a stack of graphite spheres in which the coated fuel particles are embedded. A second line of development, initiated in Europe but carried forward in the United States, is the prismatic core in which vertical replaceable graphite prisms containing graphite fuel rods (in which the coated particles are embedded) and coolant passages make up the core. Typically, core power densities range between 5 and 10 MW/m3 with helium coolant outlet temperatures up to 1000°C. A number of prototype HTGR plants have been built to demonstrate both the pebble-bed and the prismatic-core concepts, although no commercial power plant is currently in operation.

2.4.3   Pressurized-Water Reactors

By far the most common civilian reactor is the pressurized-water reactor (PWR). Reactors of this type were originally developed to drive nuclear submarines. The PWR circuit is illustrated schematically in Figure 2.7. Water at typically 150 bars (2200 psia) is pumped into a pressure vessel, which contains the reactor core. The water passes downward through an annulus between the reactor core and the pressure vessel and then flows up over the fuel elements. It then leaves through a series of pipes, which pass to the stream generator. The light-water coolant also acts as the moderator for this reactor. The absorption of neutrons by the light water (as described in Chapter 1) necessitates a significant enrichment of the fuel to 3.2% ^{235}U (~4.5 times the concentration in natural uranium).

In the steam generator, the hot water from the reactor passes through vertical U-tubes (Figure 2.9), and water at lower pressure is fed into the steam generator shell and contacts the outside of the U-tubes. Steam is generated at approximately 70 bars (1000 psia) and passes from the steam generator into the

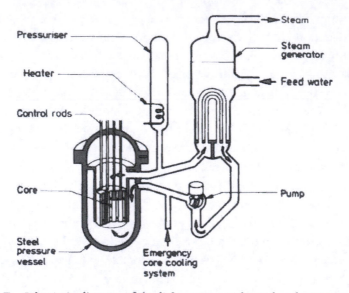

Figure 2.7: Schematic diagram of the light water–moderated and water-cooled pressurized-water reactor (PWR).

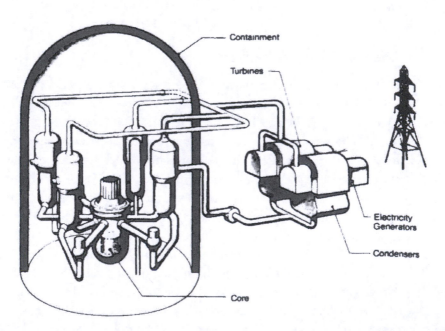

Figure 2.8: A typical four-loop PWR station.

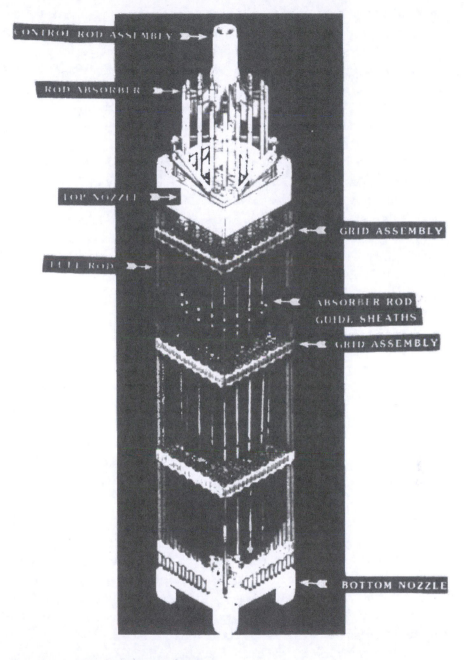

Figure 2.9: PWR fuel element design.

turbine and from there to the condenser, the condensate being returned to the steam generator via feed preheaters. Figure 2.7 illustrates one complete coolant loop; PWRs typically have two, three, or four such loops per reactor vessel. A typical four-loop PWR is illustrated in Figure 2.8. The fuel elements in a PWR are illustrated in Figure 2.9; the fuel is in the form of uranium oxide pellets mounted in a 12-ft-long tube made of a zirconium alloy (Zircaloy). The tubes are usually mounted in separate bundles of 17 rows of 17 tubes, with some pins omitted to allow passage of control rods into the core.

In 1993 there were 243 operating civilian PWR power reactors in the world and 33 under construction. Although the steam cycle efficiency of a PWR is relatively low (32%), its capital cost may be considerably less than that of an AGR. The main reason for this is the great reduction in core size made possible by the enormous increase in volumetric power density and core rating, as shown in Table 2.3. Another factor contributing to the low capital cost is the fact that much of the PWR can be constructed off-site under factory conditions.

Because of the high rate of heat generated per unit mass of fuel (*fuel rating*), the response of a PWR to changes in operating conditions is much more rapid than that of an AGR. It has been argued that this is a negative safety factor. Even when the reactor is shut down, the level of decay heat is such that the fuel must always be kept covered with water. We shall discuss these safety features in Chapters 5 and 6. Pressurized-water reactors have experienced problems with steam generators, which have failed due to corrosion on the secondary (steam-generating) side. Reactors are often more susceptible to problems outside the core than in it. Although it is now believed that design improvements can prevent these corrosion problems, most existing reactors are still prone to them. This is not a major safety issue, but it does limit their performance.

2.4.4 Boiling-Water Reactors

The boiling-water reactor (BWR) differs from the PWR in that it generates steam directly within the core and does not have a separate steam generator. The system is illustrated in Figure 2.10*a*. Water at a pressure of about 70 bars (1000 psia) is passed through the core, and about 10% of it is converted to steam. The steam is then separated in the region above the core, the water being returned to the bottom of the core via the circulation pump and the steam passing from the top of the vessel to the steam turbine. The steam from the turbine is passed through a condenser, and the condensate is returned to the reactor vessel as shown in Figure 2.10a. The core power densities in a BWR are about half those in a PWR (though still much higher than those in gas reactors). The fuel elements consist of 12-ft-long bundles of Zircaloy-canned UO_2 pellet fuel with an enrichment similar to that in PWR. Each bundle of fuel is contained within a square channel constructed of Zircaloy, as illustrated in Figure 2.10b.

The advantage of the BWR is the elimination of the steam generator, which has been one of the most troublesome features of the PWR. However, in the PWR the coolant passing through the reactor is contained within the reactor/steam generator/circulator circuit. In the BWR the coolant also passes

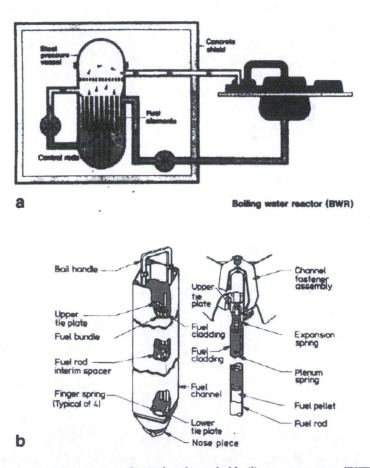

Boiling water reactor (BWR)

Figure 2.10: (a) Light water–moderated and –cooled boiling-water reactor (BWR); the fuel is enriched uranium oxide; (b) fuel bundle contained within Zircaloy channel.

through the steam turbine and the condenser. Corrosion products and in-leakage from the turbine and condenser are passed to the reactor, where they may be activated by the reactor neutrons to produce radioactive isotopes, which circulate around the system. Also entering the coolant stream are small amounts of radioactive substances leaking from damaged fuel elements, including the rare gases xenon and krypton. These find their way into the inert gas removal system in the condenser. Thus, the reactor must be operated with many of the external components maintained under radioactive conditions, which is not the case with the PWR. Consequently, BWRs give somewhat higher (though carefully limited) radiation doses to their operators. Another problem with existing BWRs has been cracking of the stainless steel pipework due to corrosion under the highly stressed conditions. This is similar to the steam generator problems in PWRs in that it can be cured by using a different design approach (i.e., using stress corrosion–resistant material), but many existing plants will continue to be susceptible to it.

2.4.5 Natural Uranium Heavy Water–Moderated and –Cooled Reactors

As discussed above, the U.S.-designed PWR and BWR reactors require considerable enrichment of the uranium in order to overcome the relatively high absorption of neutrons by the light-water coolant. This disadvantage can be overcome by using heavy water as a moderator and either heavy water or boiling light water as the coolant. If heavy water itself is used as the coolant, it is possible to operate with natural uranium. This is the principle adopted in the Canadian-designed CANDU (Canadian deuterium-uranium) reactors, which are illustrated in Figure 2.11.

CANDU reactors dispense with the massive thick-walled pressure vessel used in PWRs and BWRs; instead, the fuel elements are placed in horizontal pressure tubes constructed from zirconium alloy. These pressure tubes pass through a calandria filled with heavy water at low pressure and temperature. In the CANDU reactor, heavy-water coolant is also passed over the fuel elements at a pressure of approximately 90 bars (1400 psia). It then passes to a steam generator, which is very similar to that used in the PWR (see Figure 2.7). It should be noted that CANDU reactors have not experienced the same steam generator problems as the PWRs, possibly because of the lower operating temperature on the primary side. The fuel elements consist of bundles of natural

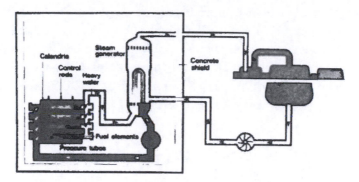

Pressurised heavy water reactor (CANDU)

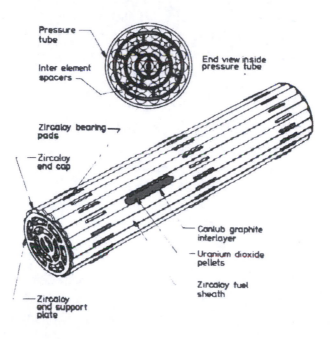

Figure 2.11: (*a*) Heavy water–moderated and –cooled CANDU reactor; the fuel is natural uranium oxide; (*b*) end and side views of a horizontal pressure tube.

UO_2 pellets clad in zirconium alloy cans; individual bundles are about 50 cm long, and about 12 such bundles are placed in each pressure tube. The average volumetric power density in a CANDU core is approximately one-tenth that in a PWR (since the moderator volume is taken into account in calculating the average volumetric power density) and nearly four times that in an AGR.

However, the fuel rating is comparable to that in a PWR. Furthermore, the fuel is very much cheaper since natural uranium can be used. Although the CANDU has operated with remarkable success, difficulties have been experienced with hydriding of the zirconium alloy pressure tubes, necessitating their replacement in some cases. Even though it has a lower fuel cost, CANDU needs considerable amounts of expensive heavy water, which makes its capital cost high.

2.4.6   Boiling-Water, Graphite-Moderated Direct- Cycle Reactor (RBMK)

The RBMK reactor is a direct-cycle boiling-water pressure tube, graphite-moderated reactor developed from Russia's first nuclear power plant, commissioned in 1954. The concept is unique to the former Soviet Union, and it was this type of reactor that was involved in the very serious nuclear accident at Chernobyl in the Ukraine in April 1986. This accident is described in Chapter 5. To aid this later description, the RBMK is covered in rather more detail than the other reactor concepts.

Figures 2.12 and 2.13 show the main elements of the reactor. The reactor core is 12 m (40 ft) in diameter and 7 m (23 ft) high, and is built up from graphite blocks (**A** in Figure 2.13) penetrated by vertical channels (**B**) and containing a zirconium alloy (Ar + 2.5% Nb) pressure tube 88 mm (3.5 in) in internal diameter and 4 mm thick. For a 1000-MW(e) reactor there are 1663 channels. Each channel contains two fuel assemblies each 3640 mm (12 ft) long, held together by a central tie rod suspended from a plug at the top of the channel.

The fuel assemblies consist of 18 pin clusters, each pin in the form of enriched (2%) uranium dioxide pellets encased in zirconium alloy tubing (13.6 mm in outside diameter. x 0.825 mm thick). The maximum power of any channel is 3.25 MW (thermal).

The fuel is cooled by boiling light water at 70-bar (1000-psia) pressure. The water enters the channel at 270°C, and the "quality" (fraction of the total mass flow that is steam) of the existing steam-water mixture is on average 14% (20% maximum).

Two separate identical coolant loops are provided. Each loop consists of two steam drums (**C**) (to which the riser pipes from the fuel channels carry the steam-water mixture) and four primary circulating pumps (**D**) (three are normally operational and one standby).

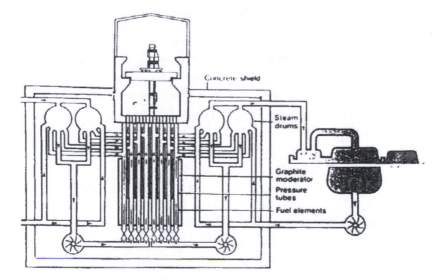

Figure 2.12: Boiling-light-water, graphite-moderated reactor (RBMK, USSR).

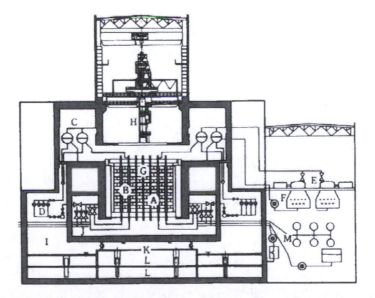

Figure 2.13: Outline diagram of the RBMK: A, Graphite blocks; B, Vertical channels; C, Steam drums; D, Circulating pumps; E, Turbine generators; F, Feed pumps; G, Absorber rods; H, Refueling machine; I, Circulating pump compartment; J, Distribution pipework; K, Surface condenser; L, Pressure supression pools; M, Emergency core cooling.

The dry steam from the steam drums passes to one of two 3000-rpm 500-MW(e) turbine generators (E). The very low pressure steam leaving the condensers is condensed in tubular condensers, and the condensate is returned to the steam drums via purifying systems and electrically driven feed pumps (F).

About 5% of the energy of the fission is dissipated in the graphite structure as a result of the slowing down of neutrons and of gamma heating. This heat is transferred to the fuel channels by conduction and radiation via a series of "piston ring"–type graphite rings that permit good thermal contact between the pressure tube and the graphite blocks while also permitting small dimensional changes. The maximum temperature of the graphite is 700°C. To improve the thermal contact and to prevent graphite oxidation, the graphite structure is enclosed in a thin-walled steel jacket through which a gas (helium-nitrogen mixture) is slowly circulated.

Perhaps the most important characteristic of the RBMK reactor is that, as originally designed, it has a positive void coefficient. This can be explained in simple terms by recognizing that if the power from the fuel increases or the flow of water decreases (or both), the amount of steam in the fuel channel increases and the density of the coolant decreases.

The term *positive void coefficient* is reactor physicist's jargon for the fact that reducing coolant density results in an increase in neutron population (light water is a strong absorber of neutrons) and hence in an increase of reactor power. However, as the power increases so too does the fuel temperature, and this has the effect of reducing the neutron population (*negative fuel coefficient*). The net effect of the positive void coefficient and the negative fuel coefficient clearly depends on the power level.

At normal full-power operating conditions the fuel temperature coefficient dominates and the net effect, termed the *power coefficient*, is negative. However, below about 20% of full power because of the lower fuel temperatures the power coefficient becomes positive. For this reason restrictions were placed on operation below 20% power.

As we shall see in Chapter 5, this fundamental design shortcoming was *the* critical factor of the accident at Chernobyl.

In short, at lower power an increase in power or a reduction of flow leads to increased boiling and further increases in power and hence to the potential for an unstable situation. As a result, RBMK requires a complex, rapidly responding control system to cope with this positive feedback.

Channels for the control and shutdown rods and for the in-core flux instrumentation pass through vertical holes in the graphite blocks. Radial flux monitors are provided in over 100 channels, and axial flux profiles are monitored in 12 channels.

The system for reactor control and protection uses 211 solid absorber rods (**G** in Figure 2.13). The rods are divided functionally as follows (Figure 2.14):

- 163 manually operated rods of which 139 are control rods (RR) for radial power shaping and 24 are dedicated to emergency protection (AZ).
- 12 rods for automatic regulation of average power (3 groups of designated AR1, AR2, and AR3, respectively).
- 12 rods for automatic regulation of local power (LAR).
- 24 shorted absorber rods (USP) for axial flux profling.

The manual control rods (RR), the automatically operated rods (AR), and the emergency shutdown rods (AZ) are distributed uniformly throughout the core in six groups of 30–36 rods. The control system includes subsystems for local automatic control (LAR) and local emergency protection (LAZ). All rods except the shortened absorber rods are withdrawn and inserted from above.

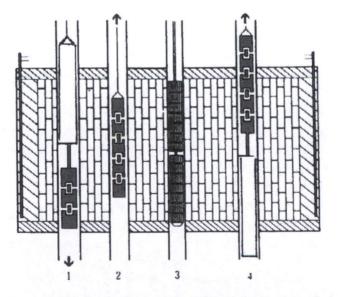

Figure 2.14: Diagram of the different control rods, "followers," and fuel assembly for the RBMK: 1, Shortened absorber rod; 2, Automatic control rod; 3, Fuel assembly; 4, Manual control rod and emergency shutdown rod.   ° followers • absorbers

The emergency shutoff rods are motor-driven at a speed of insertion of 0.4 m/s. Full insertion takes 15–20 seconds. The shorter absorber rods are introduced from below the core. The control rod channels are the same diameter as the fuel channels (88 mm) and are cooled by a separate water circuit. At the end of each rod are a number of articulated elements that do not contain neutron-absorbing material. As the rod is withdrawn these "followers" prevent water from occupying the space vacated by the absorber.

The control system is arranged to operate over the following power ranges:

1. From subcritical to 0.5% power, manual operation was used.
2. From 0.5% to 10% power, automatic regulation of overall power was performed using one of the sets of four rods designated for this purpose (i.e., AR3).
3. From 10% to 100% of the working range, overall automatic regulation was carried out using control rod groups AR1 and AR2.
4. From 10% to 100% power, local automatic regulation (LAR) was also invoked.

The reactor is "tripped" (i.e., switched off completely) only for a specific number of faults, e.g., loss of off-site (station) power, both turbines tripped, loss of three main circulating pumps, 50% loss of feed water, low steam drum water level, and high neutron flux.

For all other faults the reactor power is set back to some lower level consistent with the fault's consequence for the reactor (e.g., on loss of one circulating pump to 80% full power, trip of single turbine to 50% full power).

The RBMK reactors are designed to be refueled at full load, and Figure 2.13 shows the refueling machine (H) operating from the gantry running the length of the charge hall.

The primary coolant system is housed in a series of compartments that act as the containment in the event of an accident. Separate compartments house the primary circulation pumps (I), the coolant inlet headers and distribution pipework (J), and the reactor vault.

Each compartment is designed to withstand a pressure of 4.5 bars and is equipped with sealed electrical and mechanical penetrations and isolation valves on piping. The compartments are connected to one another and to a surface condenser tunnel (K in Figure 2.13) as well as to two pools of water ("pressure suppression pools," L) to condense the escaping steam and lower the pressure.

The steam drums (C) are housed in separate compartments on either side of the charge hall, but these are not pressure-tight compartments because of the large number of joints in the charge hall floor needed for refueling that provide a leak path between the steam drum compartments and the charge hall.

The RBMK reactor is equipped with an emergency core cooling system (M) that feeds both coolant and consists of

1. a fast-acting flooding system that automatically injects cold water into the damaged part of the reactor from two sets of gas-pressurized tanks holding enough water to cool the core for the first 3 minutes of a major loss of coolant accident. This system is supported with flow from the main feed pumps.
2. an active system of three pumps taking water from the condensate system after the pressurized tanks have emptied. These pumps are driven by three standby diesel generators that can be started within 2–3 minutes.
3. an active recirculating cooling system that consists of six pumps drawing water from the upper suppression pool through heat exchangers feeding the damaged part of the reactor and also driven by the diesels.

The emergency core cooling system is triggered by the coincidence of a high-pressure signal from any of the containment compartments and a low-level signal from the steam drums.

As a consequence of the accident at Chernobyl a number of modifications have been carried out on other RBMK units. The control rod design has been improved and the rate at which the rods can be inserted into the core has been increased. Automatic shutdown systems have been fitted to prevent the reactor from being operated continuously below 20% full power. The problem of the positive void coefficient has been reduced by fitting fixed neutron absorbers. The main influence of this measure is to alter the balance between absorption of neutrons in fixed absorbers and the variable absorption in the steam-water coolant. To compensate for these measures the enrichment of the fuel has been increased from 2.0% to 2.4% U-235.

2.5 FAST REACTORS

2.5.1. Liquid Metal–Cooled Fast Breeder Reactors

The most prevalent design for a fast reactor system is that employing sodium as the coolant. The advantages of liquid sodium in cooling reactors are discussed in Chapter 3. Briefly, sodium is an excellent heat transfer agent and can cope with the very high volumetric power densities encountered in reactors of this type (typically five times those of a PWR; see Table 2.3). The sodium-cooled fast reactor, which Is illustrated schematically in Figure 2.15, consists of a pool of sodium contained in a primary vessel in which the core is submerged.

Sodium is pumped through the core (the pumps being submerged in the sodium pool, as illustrated). The hot sodium then passes through an intermediate heat exchanger, where heat is transferred from the primary coolant to a secondary sodium stream; the secondary stream passes through the steam generator, where steam is raised for electricity generation. In contrast to the AGR and PWR, this reactor has three heat transfer stages: from the fuel elements to the primary sodium coolant, between the primary sodium coolant and a secondary coolant, and between this secondary coolant and evaporating water in the steam generator. This somewhat complex system ensures that the primary coolant stays in the primary vessel and that any radioactive substances in the primary vessel are not transferred to the steam generator, where the potential exists for a chemical interaction between the sodium and the water (due to minute leakages).

Since the reactor utilizes fast neutrons, there is no moderator. The layout of the U.K. 250-MW (electrical) prototype sodium-cooled fast reactor (PFR) is shown in Figure 2.16. A similar prototype (Phenix) has been operated in France.

A much larger commercial-sized [1200-MW(e)] fast reactor, Superphenix, has been built in France and was commissioned in 1986. European utilities, design and construction companies, and research and development organizations have

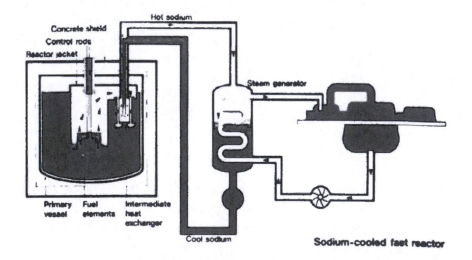

Figure 2.15: Schematic diagram of the sodium-cooled fast reactor.

collaborated on an advanced design known as the European fast reactor (EFR) with an electrical output of 1450 MW(e). The layout of the proposed EFR is illustrated in Figure 2.17. The fuel is in the form of pellets of mixed plutonium

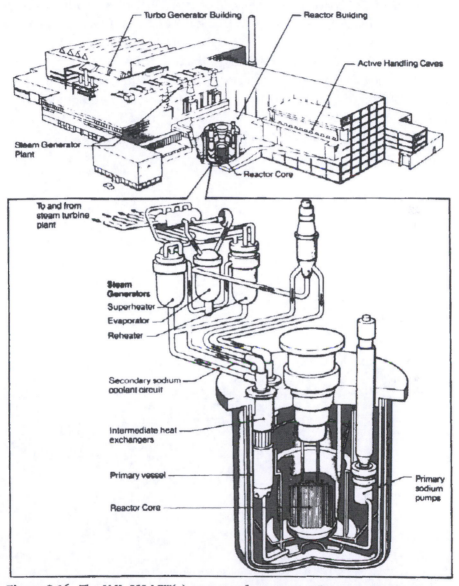

Figure 2.16: The U.K. 250-MW(e) prototype fast reactor at Dounreay, Scotland.

and uranium oxides (20–25% Pu O$_2$) clad in austenitic or nimonic alloy steel tubes as illustrated in Figure 2.18. Each fuel element consists of 331 pins, each 8.2 mm in diameter with an active core length of about 1 m. The core power density is about 5 times that in a PWR and 1000 times that in a Magnox reactor.

Sodium-cooled fast reactors have been operated in the United Kingdom, the United States, France, the former Soviet Union, and Japan. In recent years, the commercial development of fast reactors has slowed down. Problems have been encountered in the steam generators in fast reactors, where it has not always been possible to meet the requirement for complete watertightness of the tubes. However, the sodium-cooled fast reactor has some inherent safety features that may make it very attractive, despite its very high power density. We shall discuss these in Chapters 5 and 6.

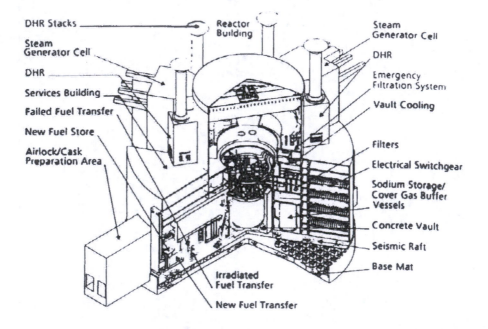

Figure 2.17: Proposed design for the European fast reactor.

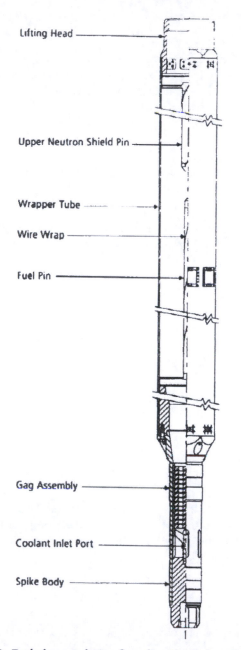

Figure 2.18: Fuel element design for a liquid metal–cooled fast breeder reactor.

EXAMPLES AND PROBLEMS

1 Power increase following increase in reactivity

Example: A sudden increase in reactivity of a water reactor 1% beyond prompt criti-cality occurs. The neutron lifetime is 10^{-4} s. What is the increase in reactor power after 1/100 s? What processes are available to terminate the transient?

Solution: The reactor power increased by a factor of $1.01^{100} = 2.7$ in 0.01 s. If such a re-activity increase is to be terminated before melting of the fuel occurs, then steam bub-bles must appear within a few hundredths of a second to expel the moderator and terminate the fission reaction.

Problem: What increase in reactivity would be required to increase the power of a water reactor by a factor of 2 in 0.01 s, assuming a neutron lifetime of 10^{-4}s?

2 Decay heat removal

Example: A 4000-MW(t) PWR has been taken out of service. Use the data given in Table 2.2 to estimate the rate of decay heat generation after 1000 h and 1 year from shutdown.

Solution: From Table 2.2 we see that after 1000 h the decay heat rate is 0.11% of the full-power rate. Thus the decay heat generation rate after 1000 h is

$$\frac{4000 \times 0.11}{100} = 4.4 \, \text{MW}$$

Similarly, after 1 year, 0.023% of full power is emitted as decay heat, giving the follow-ing value for decay heat generation:

$$\frac{4000 \times 0.023}{100} = 0.92 \, \text{MW}$$

Problem: Assume that the shut-down reactor in the example is cooled by residual heat removal (RHR) water at 20°C. Calculate the RHR water flows required after 1000 h and 1 year if the rise in water temperature is to be restricted to 20°C.

3 Fuel investment in thermal reactors

Example: Using the data in Table 2.3, estimate the investment of enriched fuel that would be required for a 10-GW(e) program of AGRs and PWRs, respectively.

Solution: With a figure of 11 MW(t)/tonne and a thermodynamic efficiency of 40%, the fuel required for the AGR program would be

$$\frac{10 \times 10^3}{11 \times 0.4} = 227.3 \, \text{tonne}$$

Similarly, assuming a thermodynamic efficiency of 32%, the fuel required for the PWR program would be

$$\frac{10 \times 10^3}{38.8 \times 0.32} = 80.5 \, \text{tonne}$$

Problem: Assuming that the alternative programs were for 1000-MW(e) reactors, use the data from Table 2.3 to estimate the core volumes required for the AGR and PWR reactor choices, respectively. Calculate the diameters of equivalent spheres required to contain these respective volumes.

BIBLIOGRAPHY

Dent, K.H., et al. (1982). "Status of Gas Cooled Reactors in the UK." In *Gas-Cooled Reactors Today,* Proceedings of a Conference, Bristol, September 20–24, 1982, vol. 3, 247–58. British Nuclear Energy Society, London, 830 pp.

Duderstadt, J.J. (1979). *Nuclear Power.* Marcel Dekker, New York.

Haywood, R.W. (1975). *Analysis of Engineering Cycles.* Pergamon, Elmsford, N.Y.

Hirsch, P.B. (1990). *The Fast-Neutron Breeder Fission Reactor.* The Royal Society, London.

International Atomic Energy Agency (1986). *Nuclear Power Reactors in the World.* IAEA, Vienna, April 1994.

International Atomic Energy Agency (1986). *The Accident at Chernobyl Nuclear Power Plant and Its Consequences.* Information for the IAEA Experts Meeting, August 25–29, 1986. Compiled by the USSR State Committee on the Utilization of Atomic Energy.

Knief, R.A. (1992). *Nuclear Energy Technology: Theory and Practice of Commercial Nuclear Power,* 2d ed. Hemisphere/Taylor and Francis, Washington, D.C., 770 pp.

McIntyre, H.C. (1975). "Natural Uranium Heavy Water Reactors." *Sci. Am.* 233 (4): 17–27.

National Nuclear Corporation (1986). *The Russian Graphite Moderated Channel Tube Reactor.* Report of a Critical Assessment following a Visit to Leningrad RBMK Station, March 1976 (republished May 1986).

Patterson, W.C. (1983). *Nuclear Power,* 2d ed. Penguin, Harmondsworth, U.K., 256 pp.

Weisman, J. (1977). *Elements of Nuclear Reactor.* Elsevier, New York.

Winterton, R.H.S. (1981). *The Thermal Design of Nuclear Reactors,* Pergamon, Elmsford, N.Y.

3

Cooling Reactors

3.1 INTRODUCTION

As we saw in Chapter 2, a variety of liquids and gases have been used to cool nuclear reactors. The present chapter introduces some of the desired general features of a reactor coolant and discusses the actual processes of heat transfer from the fuel elements to the primary coolants and from the primary coolants to the steam generation system. It also reviews the various types of coolant (gaseous, liquid, and boiling) and concludes by giving some examples of the engineering features of cooling circuits used in various types of reactor.

3.2 GENERAL FEATURES OF A REACTOR COOLANT

The general features that make a particular fluid (gas or liquid) attractive as a reactor coolant are as follows.

1. *High specific heat.* Suppose we have a nuclear reactor that is generating heat at a rate of Q watts. Coolant at a flow rate W (kilograms per second) is passed to the reactor, entering the core at temperature T_{in} and leaving the core at temperature T_{out}. From the first law of thermodynamics (see Section 1.1.1), these quantities are related by the equation $Q = WC_p (T_{out} - T_{in})$, where C_p is the *specific heat* or *specific heat capacity* of the fluid. The specific heat is the amount of heat required to heat 1 kg of a substance by 1 K (1°C) and thus has the units joules per kilogram per kelvin. In designing reactors it is important to prevent excessive temperatures within the core, in order to avoid damaging the fuel and the core construction materials. The above equation indicates that this can be accomplished in two ways for a given inlet temperature of the coolant. First, the flow rate W can be so high that the outlet temperature is not too much higher than the inlet temperature, irrespective of the value of C_p. Second, a fluid can be chosen that has a high value of C_p which will also limit the outlet temperature. Of course, the outlet temperatures cannot be too low, or the reactor will not be able to generate steam efficiently, as explained in Chapter 1. Also, with high flow rates

significant amounts of power are needed to pump the coolant, and this is power that is not available as electricity to the customer.

A special case is that in which the coolant is in the form of a boiling liquid. Here, heat can be absorbed by the coolant at its boiling point with no change in temperature and can be used to convert the liquid into vapor. The amount of heat required to convert one unit mass of liquid to vapor is called the *latent heat* of vaporization (joules per kilogram). The boiling-fluid coolant is often also used as the working fluid in the turbine (e.g., steam generated from a boiling-water coolant in a reactor is used in a steam turbine). For the reasons discussed in Chapter 1, the higher the boiling point of the fluid the higher the thermodynamic efficiency. Since boiling point increases with pressure, the boiling-coolant system should be operated at the highest practicable pressure. However, the higher the pressure, the more expensive the system, and there is a trade-off between increased capital cost and increased thermodynamic efficiency.

2. *High rates of heat transfer.* The rate at which heat can be transferred from the fuel elements to the coolant is determined by a number of factors, which are discussed in more detail in Section 3.3. One of the parameters is the *thermal conductivity* of the fluid, which is the constant of proportionality between the rate at which heat is transferred through a static volume of fluid and the *temperature gradient*, i.e., the rate at which temperature is changing per unit length. Liquid metal coolants have high thermal conductivity, whereas gaseous coolants have relatively low thermal conductivity.

3. *Good nuclear properties.* For all reactors, it is important that the coolants should have low neutron absorption. As explained in Chapter 2, any neutron absorption by the coolant and structure reduces the number of neutrons available for the fission reaction. The neutrons should not react appreciably with the coolant to form radioactive isotopes. Excess radioactivity in the circulating system increases operational difficulties, as mentioned in Chapter 2. If the coolant is also acting as the moderator, good moderation properties are required (the processes of moderation were explained in Chapter 1). In fast reactors, of course, it is important that the coolant not moderate the neutrons, since unmoderated (fast) neutrons are required in the reaction.

4. *Well-defined phase state.* It is preferable for the coolant to have the same phase state (i.e., liquids remain as liquids and gases remain as gases) during both normal and accident conditions. To achieve this in the case of liquids, a high boiling point is desirable to avoid changes of phase if the liquid is overheated. A high boiling point also has the advantage of minimizing the pressure required to operate at a certain temperature level and of achieving high thermodynamic efficiency.

5. *Cost and availability.* Since the inventory of coolant in typical reactor systems is quite high (hundreds of tons), it is important that the cost be minimized. Also, coolants may leak from reactor circuits, and this can be a

significant cost in some cases. The ideal coolant should also be freely available in a sufficiently pure form for use in the reactor circuit.

6. *Compatibility.* It is obviously axiomatic that the coolant should be compatible with the reactor circuit and not corrode it, even under the conditions of high radiation flux that occur in the core.

7. *Ease of pumping.* Fluids of low viscosity require much less pumping power to circulate them around the reactor circuit than do fluids of high viscosity. The viscosity of a fluid is related to its temperature, that of liquids decreasing with increasing temperature and that of gases increasing with increasing temperature. The viscosity of a fluid is indicated by the symbol μ.

No practical fluid meets all of these requirements. All known coolants have one or more disadvantages. The thermodynamic and heat transfer characteristics of a coolant can be compared conveniently by using a parameter called the *figure of merit*, which derives from the heat transfer processes and the associated pumping power required. The figure of merit F is defined as

$$F = \frac{C_p^{2.8} \varrho^2}{\mu^{0.2}}$$

where C_p is the specific heat, ϱ the fluid density (kilograms per cubic meter), and μ the viscosity. The rather peculiar-looking powers appearing in this equation result from the empirical correlations used to predict the pumping power and the heat transfer rates.

There are relatively few practical choices for reactor coolants. The ones mainly used are listed in Table 3.1, which shows their density, viscosity, specific heat, thermal conductivity, and figure of merit value. In terms of figure of merit, ordinary water is outstanding. However, it has three main disadvantages: its low boiling point, which requires operation at high pressure in order to reach even moderate thermodynamic efficiencies; its neutron absorption; and its corrosion properties. The latter two disadvantages require enrichment of the fuel and special containment materials, respectively.

3.3   PRINCIPLES OF HEAT TRANSFER

In discussing heat transfer processes, it is usual to define a *heat flux* $\dot{q}$ from a surface, which is the rate of heat flow per unit surface area per unit time and has units joules per square meter per second or watts per square meter (a watt is a joule per second). The heat flux is commonly related to the *temperature dif-*

Table 3.1 Physical Properties of Reactor Coolants

Coolant	Melting Point (°C)	Boiling Point (°C)	PHYSICAL PROPERTIES GIVEN AT		Density (kg/m^3)	Viscosity [Ns/m^2 (x 10^6)]	Specific Heat (kJ/kg °C)	Thermal Conductivity (W/m °C)	Relative Figure of Merit[a]	Macroscopic Thermal Neutron Absorption Cross Section (cm^{-1})
			T (°C)	p (atm)						
Light water	0	100	270	54	767	102	5.14	0.059	53	0.017
Heavy water	4	101	270	54	845	113	5.27	0.049	67	2.8 x 10^{-5}
Sodium	98	883	550	1	817	230	1.26	6.1	1	0.011
p-Terphenyl	213	427	400	1	880	100	2.2	0.013	6.5	0.008
Helium	−272	−269	450	40	3.08	36	5.2	0.028	1.1 x 10^{-3}	2 x 20^{-9}
Carbon dioxide	−57	−78	450	40	29.5	30	1.2	0.07	1.7 x 10^{-3}	10^{-7}

Source: Etherington (1958).

[a] Value of $C_p 2.8 \rho^2 / \mu^{0.2}$ divided by that for sodium (hence value for sodium is unity).

ference or *temperature driving force* ΔT by the simple equation:

$$\dot{q} = b \, \Delta T$$

where *b* is a constant of proportionality commonly referred to as the *heat transfer coefficient*. The temperature difference ΔT is defined as the difference between the fuel element surface temperature T_W and the bulk coolant temperature T_B:

$$\Delta T = T_W - T_B$$

The temperature of the fluid is not uniform across the channel; the fluid adjacent to the wall is at the wall temperature. The bulk temperature T_B is defined as the fluid temperature that would be obtained if the fluid were totally mixed within the channel. Figure 3.1 shows a typical temperature distribution across the fuel and coolant in a reactor. Heat is generated in the fuel pellets and is conducted to the pellet surface, then across the gas gap between the pellet and the can, then through the can wall, and finally out to the fluid.

The heat transfer processes in the reactor must be designed to prevent the system from exceeding two main temperature limits:

1. *Maximum temperature of the fuel.* If the fuel is made from uranium metal, its maximum temperature is around 650°C, where volume swelling occurs due to a crystal structure change in the metal. For uranium oxide fuel, the maximum temperature is around 2800°C, the melting point of the oxide. Despite its much lower maximum temperature, metal fuel may release heat from its surface at a higher rate than oxide fuel because of its much higher thermal conductivity. However, in modern reactors metal fuel is rarely used, since it undergoes chemical reaction with the coolant if the cladding is ruptured.

2. *Maximum cladding temperature.* The temperature of the cladding material is often the limiting factor. For instance, the commonly used Zircaloy cladding rapidly corrodes if its temperature is greater than about 500"C, and it reacts exothermically (i.e., generates heat, which can promote further reaction) with steam to form hydrogen at temperatures above 1000°C. Stainless steel cladding is used in AGRs and liquid metal–cooled fast reactors; it is compatible with carbon dioxide and sodium at normal operating conditions (700–750°C) but oxidizes rapidly at higher temperature, the short-term absolute limit being the stainless steel melting point of about 1400°C.

In practice, it is not feasible to design a nuclear reactor system to work close to these maximum temperatures, since a margin must be provided for abnormal or accident conditions. Typical maximum cladding temperatures for steady operation of various reactor systems are as follows:

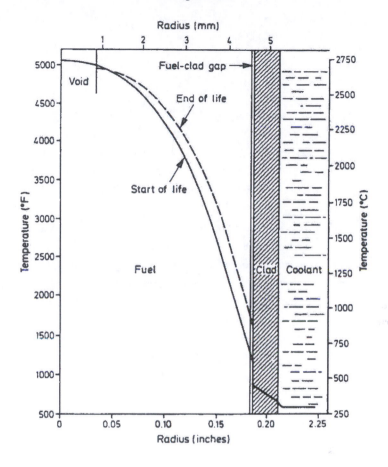

Figure 3.1: Typical fuel pin temperature profile (PWR fuel).

Magnesium alloy cladding (Magnox)	450°C
AGR stainless steel cladding	750°C
Pressurized-water reactor	320°C
Boiling-water reactor	300°C
Sodium-cooled fast reactor	750°C

The heat transfer coefficient b depends on the physical properties of the fluid, increasing with increasing fluid thermal conductivity, decreasing fluid viscosity, and increasing fluid density. It is also a strong function of the fluid velocity. Typical values of b for reactor coolants at the usual ranges of velocity are as follows:

Water	30,000 W/m² °C
Boiling water	60,000 W/m² °C
High-pressure carbon dioxide	1,000 W/m² °C
Liquid sodium	55,000 W/m² °C

In a pressurized-water reactor the heat flux $\dot{q}$ is typically around 1.5 million W/m², giving a cladding-to-fluid temperature difference of about 50°C. In a liquid metal–cooled fast reactor, the heat flux might be typically 2 million W/m², giving a cladding-to-fluid temperature difference of about 35°C. Similarly, in a boiling-water reactor, a typical heat flux is 1 million W/m², giving a temperature difference of around 15°C.

The values given above for heat transfer coefficients are those appropriate for smooth, plain surfaces. The values for carbon dioxide are very much lower than those for water and sodium. This means that the temperature difference would be unacceptably high, or the power output unacceptably low, for gas-cooled systems. It is thus necessary to *enhance* the heat transfer in some way in these systems. In Magnox reactors this is done by using external fins, typically of the form illustrated in Figure 2.4 and in more detail in Figure 3.2. The fins on the surface increase the area of cladding in contact with the gas, thus increasing the heat transfer rate for a given amount of fuel. The fins also promote intense

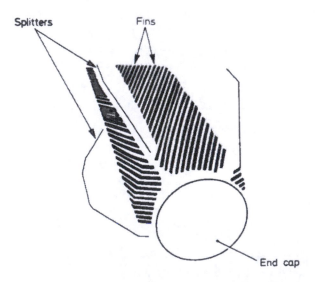

Figure 3.2: End view of a Magnox fuel can with herringbone pattern of fins.

mixing of the gas, which also aids the heat transfer. By using external fins, the heat transfer rate is increased above that for a plain can by a factor of 5 to 6.

In the advanced gas-cooled reactor (AGR), enhancement of gas-phase heat transfer is achieved by quite different means. The can is machined to produce rectangular ribs on the surface as illustrated in Figure 3.3. These ribs add only slightly to the total surface area of the cladding, but they enhance the heat transfer coefficient by a factor of typically 2.5. By interrupting the flow of the hot gas along the surface and causing the hot gas to be mixed with the cooler gas in the bulk flow, they help bring the cooler gas to the surface, enhancing the heat transfer rate. However, this enhancement of heat transfer is achieved at the expense of increasing the frictional resistance to gas flow through the system, thus requiring more power to drive the circulators.

In nuclear electricity generation, it is necessary to boil water in order to produce steam. In the boiling-water reactor, this is done directly in the reactor core (see Figure 2.10) In the other reactor types discussed in Chapter 2, boiling occurs in a separate steam generator, which is heated by the primary coolant: water (PWR), carbon dioxide (AGR), or sodium (fast reactor).

The phenomenon of boiling is encountered frequently in everyday life. Most British families have an electric kettle to produce boiling water for domestic purposes. In such kettles, bubbles of steam are produced at the heating element surface and rise through the water, initially condensing but later escaping from

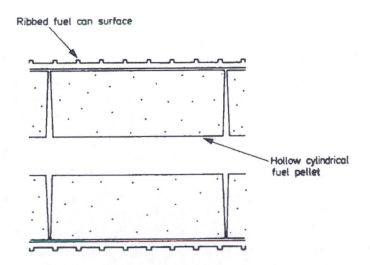

Figure 3.3: Longitudinal section through an AGR fuel rod.

the surface of the water and out through the kettle spout, at which time most people remember to switch off the kettle. In a typical kettle, the heat flux would be around 150,000 W/m². For such a domestic kettle, the heat transfer coefficient would be around 10,000 W/m² °C, giving a temperature difference between the surface of the element and boiling water of about 15°C. The electric kettle provides a useful analogy in discussing safety issues and accident conditions in Chapter 4. Note that the heat transfer coefficient for a typical domestic kettle is approximately one-sixth of that observed for boiling in a boiling-water reactor, because the heat transfer coefficient in boiling increases with increasing pressure and with increasing heat flux, both of which are higher in the BWR.

A further complication in the BWR is that the steam generated flows along with the remaining water, resulting in a *two-phase flow* (the two phases being water vapor and water liquid). Two-phase flows are highly complex in nature and have higher flow resistance (higher pressure drop through the reactor) than equivalent single-phase flows. The development of two-phase flow in a heated channel is illustrated for the case of a simple heated tube in Figure 3.4. At the bottom of the channel, heat transfer is to the liquid alone (i.e., a single phase). At a certain point along the channel, bubbles start to form at the wall, and we enter the *bubbly* two-phase flow regime. Initially, the bubbles are formed at the wall and condense rapidly when they move toward the center of the tube. However, when the liquid heats up to its boiling point the bubbles can no longer condense. As the flow proceeds farther up the tube, more and more of the fluid is in the form of steam. A parameter commonly used to describe the extent of evaporation is the *steam quality x*, which is the fraction of the total mass flow in the form of vapor. The quality increases along the channel as vapor is generated as a result of the transfer of heat to the fluid. When the population of bubbles is sufficiently high, they begin to coalesce and form very large bullet-shaped bubbles, which characterize the *slug* flow regime. Eventually, these slug flow bubbles all join together, and we enter the *annular* flow regime, where there is a liquid film on the heated surface with the vapor flowing in the center of the channel (Figure 3.4). The surface of this liquid film is highly disturbed by ripples and waves, and liquid is picked up from the wave tips in the form of droplets and flows with the steam.

Farther along the channel, the liquid film is gradually thinned by the process of evaporation and droplet formation and finally dries up. Here, the *drop* flow regime is entered, with the liquid phase flowing totally as droplets. The transition from the annular flow (wetted wall) to the drop flow (dry wall) region is often referred to as *dryout* or *burnout*. This is a particularly important transition since it results in a large decrease in the heat transfer coefficients. In the annular flow regime, the coefficient is typically many tens of thousands of watts per square meter per degree centigrade. Beyond the transition, in the drop flow regime, the coefficient can fall to a small fraction of this value, typically 2000 W/m² °C. This large decrease in heat transfer coefficient results in an increase in heating surface temperature if the heat flux is maintained constant. As a result, the heating surface may become unacceptably hot. It is important to avoid the dryout-burnout transition in the reactor core situation, where the heat flux is governed mainly by the neutron population. As shown in Section 2.2, the tem-

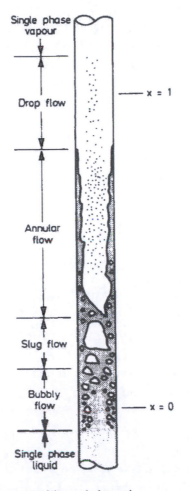

Figure 3.4: Flow patterns in a vertical heated channel.

perature of the fuel in an operating nuclear reactor is determined by the rate of heat transfer into the coolant. If the heat transfer coefficient falls by a factor of, say, 30, from 60,000 to 2,000 $W/m^2 \, °C$, then the temperature difference between the fuel and the coolant will rise by an equivalent factor, namely, from 15°C to 450°C, which would exceed the permissible operating temperature for Zircaloy cladding. It is thus very important to operate nuclear reactors under conditions at which dryout-burnout does not occur.

Referring to Figure 3.4, we see that the droplets persist for long distances beyond the dryout point. This occurs because the droplets evaporate slowly, even if the steam is heated well above the boiling point or *saturation temperature*. Heat transfer in the region beyond dryout-burnout is very important in considering accident conditions and will be discussed in Chapter 4

In contrast to the situation in the reactor core, where the heat flux is controlled by the neutron population, boiling in the steam generators of indirect-cycle reactors (AGR, PWR, fast reactor) is controlled by the temperature of the primary coolant fluid. Thus, if and when the dryout-burnout transition is traversed, the heat flux itself will decrease commensurate with the decreased heat transfer coefficient. In one design of PWR steam generator (the "once-through" steam generator design of Babcock & Wilcox), the dryout-burnout transition is deliberately traversed. This is also the case in the steam generators of the AGR and in some steam generator designs for fast reactors.

3.4 GASEOUS COOLANTS

Gaseous coolants have the great advantage of having a well-defined phase state. Unlike liquid coolants, they are not subject to a change of phase, with the resulting complicated two-phase flow problems during abnormal operating conditions. However, they have the disadvantages of a low heat capacity and low heat transfer coefficients, the latter necessitating heat transfer enhancement or low operating temperatures. A wide variety of gases have been considered for nuclear reactor cooling, but only those that have been used or have had serious evaluation will be discussed here.

3.4.1 Air

Air cooling was used in the very first generation of nuclear reactors, namely, graphite-moderated natural uranium "piles," which were built in both the

United Kingdom and the United States in the 1940s. The largest air-cooled reactors were at the U.K. Atomic Energy Authority's Windscale establishment and were designed for plutonium production. The main problem with air as a coolant is that it is an oxidant, i.e., it supports combustion. In the case of the Windscale graphite-moderated piles, there was something of a dilemma: if the pile temperature was too high, the graphite oxidized, but if the pile temperature was too low, the graphite atoms could become permanently displaced from their natural positions by neutron bombardment. At higher temperatures, the atomic vibrations are sufficient to shake them back to their normal positions. Displacement of the atoms results in energy being stored, with possible accident connotations, which we will discuss in the context of the Windscale accident in Chapter 5.

Despite the ready availability of air, its oxidizing properties rule it out as a viable coolant in modern high-temperature reactors.

3.4.2 Carbon Dioxide

In terms of its physical properties, carbon dioxide is the best available gaseous coolant, and consequently it was chosen for the large U.K. Magnox and AGR power stations. In the Magnox reactors, the graphite moderator has a maximum temperature of only about 350°C. At these temperatures CO_2 is unreactive with graphite, nor does it react with the canning material, the circuit steels, or the fuel (uranium metal). When the temperature is increased, difficulties arise because of the chemical reaction:

$$CO_2 + C \rightarrow 2CO$$

Here, the C (carbon) represents the moderator graphite blocks, and the reaction slowly removes the moderator from the reactor, decreasing the strength of the graphite core. The above reaction is induced not only by higher temperatures but also by increasing nuclear radiation.

The occurrence of the carbon dioxide-graphite reaction is a potentially very serious limitation since the structure of the core, including the alignment of the fuel channel, is dependent on the physical strength of the graphite blocks. In the case of the advanced gas-cooled reactor, there have been two approaches to solving this problem:

1. The inlet (relatively cold) carbon dioxide is fed through the moderator structure to the entrance of the fuel channels, thus keeping the moderator at a lower temperature.

2. Carbon monoxide and methane are added to the carbon dioxide to inhibit the above chemical reaction. The mechanisms by which this inhibition is achieved are complex. One mechanism is that the additives produce a thin layer of carbon on the graphite, and this carbon layer reacts sacrificially with the coolant, preventing attack on the bulk structural graphite. A difficulty here is that the carbon may, under certain circumstances, be deposited on the fuel elements themselves. As we saw in the preceding section, heat transfer from the fuel elements is critically dependent on small isolated rib roughnesses on the surface of the cladding. Smoothing out of these roughnesses by carbon deposition would negate their enhancement of heat transfer and lead to a rise in the fuel element temperature. Very precise chemical control is therefore required in AGRs.

3.4.3   Helium

Helium is one of a family of gases (which also includes argon, neon, and xenon) commonly referred to as the *inert gases* or *noble gases*. Apart from some exceptions of purely academic interest, atoms of these gases do not form compounds with other elements (hence their description as inert). Helium, which has a molecular weight of 4, is present in small quantities in the atmosphere but is more commonly derived from oil and natural gas wells.

The inert gas argon (atomic mass 40) is much more available; air contains 0.94% by volume of this gas. Unfortunately, argon is not suitable as a reactor coolant, since irradiation by neutrons causes it to form a radioactive isotope (argon-41) that decays with a half-life of 1.8 h, emitting both β and γ rays. This neutron absorption and the resultant activation of the coolant circuit are unacceptable. Helium, though more expensive than argon, is not activated in a neutron flux and is, therefore, much more suitable.

Helium has been employed in the so-called *high-temperature gas-cooled reactor* (HTR). Here, the fuel is in the form of uranium carbide clad in graphite, which acts as both the cladding material and the moderator. With helium it is possible, in principle, to operate such reactors at very high temperatures (typically in excess of 800°C) without any chemical attack on the moderator-clad. However, it is usually impossible to maintain the helium coolant in a pure state, because in an actual circuit there will be a small leakage of water vapor from the boilers, ingress of air and other materials through leaks of the circulators, and release of gases originally adsorbed on the graphite. Although the helium itself does not react with the graphite or the steel structures even at high operating temperatures, the impurities do, and this limits the temperatures that can be achieved and provides a major design problem for such reactors.

3.4.4 Steam

Steam has better thermodynamic properties as a coolant than carbon dioxide. Its high specific heat allows good heat transport with lower mass flow rates and smaller, more compact piping systems than in other gas-cooled units. This has led to a number of studies of the possibility of using steam as a reactor coolant. However, at high temperatures and pressures, steam is a highly corrosive oxidizing fluid, and stainless steels may be the only suitable construction materials for use with steam at temperatures above about 600°C.

In a conventional oil-fired or coal-fired boiler, it is normal to superheat the steam (i.e., increase its temperatures above the saturation temperature) before feeding it to the turbine. This increases the overall thermodynamic efficiency of the power generation of the cycle. In the normal nuclear boiler (e.g., the BWR), the steam is not superheated. However, there have been a number of attempts to introduce superheating in nuclear boilers and to make the nuclear reactor closer to a conventional system. In this case, the steam can be regarded as a supplementary coolant to the boiling water in the other parts of the reactor. In general, it is not economically attractive to introduce superheating in this way, mainly because it requires the use of stainless steel cans and hence an increase in the enrichment of the fuel. However, a number of plants in the former Soviet Union do employ superheating on a regular basis.

3.5 LIQUID COOLANTS

In contrast to gaseous coolants, liquid coolants may undergo a change of phase (i.e., into vapor) if their temperature rises high enough. However, they have a much higher heat capacity and their better heat transfer characteristics (discussed in Section 3.3) allow them to be operated at much higher heat fluxes than gases. A variety of liquids have been used in reactor cooling, but only water (light and heavy), organic fluids, molten salts, and liquid metals will be considered here.

3.5.1 Light Water

From the earliest days of the development of nuclear energy, reactor cooling by ordinary water (i.e., light water) has been the most commonly adopted practice. Water can be used as both coolant and moderator, as in the pressurized water reactor (PWR; see Chapter 2) or in combination with a separate moderator such as graphite or heavy water. An example of the former combination is the Russian boiling-water, graphite-moderated direct-cycle reactor (RBMK). Here, the fuel channels cooled with light water pass through pressure tubes set within the graphite core, which acts as the moderator.

Although light water is readily available, a number of problems are associated with its use:

1. It has a relatively low boiling point (100°C), and therefore the reactor must be operated at high pressure to maintain the water in the liquid state at temperatures suitable for power generation cycles. Thus, in the PWR the light water is pressurized up to 15.5 megapascals (155 bars, 2300 psia), at which its saturation temperature is 345°C. The average outlet temperature for the water from a PWR is around 320°C, although the temperature may exceed the saturation temperature (giving some local boiling) in some of the channels.

2. In the presence of a neutron flux, water decomposes slightly into its constituent elements (hydrogen and oxygen). This radiation-induced reaction can be suppressed by having an excess of hydrogen dissolved in the water, which is the system adopted in the PWR.

3. Water is actually quite a corrosive substance, reacting with the materials of the fuel elements and reactor circuit and picking up trace amounts of the variety of elements present. These elements are in the form of dissolved or suspended material and may be activated in the neutron field to give radioactive isotopes, which remain in the water or deposit around the circuit. Thus, the primary circuit of a PWR is generally rather radioactive and requires remote maintenance procedures. This activation can be minimized by very careful control of the water chemistry within the circuit and the choice of reactor materials. This subject is of great importance in the economics of the system operation.

4. As discussed in Chapter 1, light water is a fairly strong absorber of neutrons, which leads to two problems. The first is the need for extra enrichment of the fuel, and the second is that under certain circumstances, accidental removal of the coolant water from the reactor core (e.g., by replacing the water by steam in a loss-of-coolant situation) may lead to an increase in the neutron population and an increased rate of the nuclear reaction. This is not a problem if the light water also serves as a moderator (as in the PWR and BWR), since the moderator is also removed and the nuclear reaction stops. In the Russian RBMK reactor, where the main moderator is graphite, there are potential problems with a loss of cooling water leading to a reactivity increase if the fuel channels are voided—what is often referred to as a positive void coefficient.

Thus, although light water is the most widely used coolant, it clearly is not ideal, but, to be fair, neither is any other coolant.

3.5.2 Heavy Water

Heavy water (deuterium oxide, D_2O) is present to the extent of 0.016% in ordinary water. Heavy water may be separated from ordinary water by various processes, which is an expensive business requiring a very large plant. Nevertheless, heavy water has considerable merit as a reactor coolant; it has a much lower thermal neutron absorption cross section than light water, which enables reactors using heavy water as a coolant to be operated without enrichment of ^{235}U in the fuel. The most common example of such a reactor is the Canadian CANDU, which was described in Chapter 2.

With the exception of neutron absorption, heavy water has practically the same physical properties and therefore the same disadvantages as light water.

Since heavy water is a very valuable material, losses and contamination with light water must be minimized. This demands a high-integrity primary circuit, particularly in the steam generators, where light and heavy water are separated only by the heat transfer surface. In practice, an annual loss of about 2% of the heavy-water inventory seems unavoidable, probably mainly in the form of vapor escaping through leaks.

Another problem with heavy water is that in a neutron flux the component deuterium is converted, to a small but significant extent, to tritium (hydrogen-3), which is radioactive and decays to helium-3 with the emission of a β-particle. Because tritium has a relatively long half-life (12 years), tritium contamination of the environment by coolant leaks from the reactor is a problem that must be taken into account in the design.

3.5.3 Organic Fluids

In an attempt to overcome some of the disadvantages of water, particularly its low boiling point and consequent high operating pressure, reactor systems have been proposed employing various organic fluids. In practice, only one group of compounds, the *polyphenyls*, have proved sufficiently resistant to neutron radiation to be of interest. In general, these coolants are mixtures of polyphenyls chosen so that they remain liquid at room temperature. It is possible to operate these coolants as liquids in excess of 300°C at operating pressures of about 10 bars, compared with 155 bars required for water. In their pure form, these coolants are essentially noncorrosive to common reactor materials.

The big problem with organic coolants is that although they are *relatively* resistant to degradation by irradiation and thermal degradation, these processes still occur to a significant extent. Radiolysis causes the formation of hydrogen and gives rise to a breakdown phenomenon called *hydrogen embrittlement* of the fuel canning. Also, the irradiation leads to the formation of polymers (materials of very high molecular weight), which deposit as a solid on the fuel elements. Although reactors have been operated with such coolants, they have not found general acceptance in commercial systems.

3.5.4 Molten Salts

Higher operating temperatures at lower pressures can be obtained by using molten salts as coolants. Molten metal hydroxides such as sodium hydroxide (caustic soda) have been suggested. The melting point of such substances tends to be rather high, though by the use of mixtures, as of sodium and potassium hydroxides, lower melting points (typically 190°C) can be obtained. The main problem with such systems is corrosion, and this has prevented their serious application.

In the early days of the development of nuclear power, many reactor systems were suggested in which the fissile material (e.g., in the form of uranium tetrafluoride, UF4) was dissolved in a mixture of fused salts. When this mixture is passed through a vessel containing a moderator such as graphite, a fission reaction takes place, heating the uranium-containing fused salt. The fused salt is pumped from the reaction zone to a heat exchanger, where the heat is transferred to another heat transfer fluid and ultimately to a power generation system. A reactor of this type, in which the fuel is actually dissolved in the coolant, is termed a *homogeneous reactor* and has the advantage that the fuel can be reprocessed continuously. However, the corrosion and other problems associated with reactors of this type effectively rule them out, though small prototypes have been operated.

3.5.5 Liquid Metals

Molten (liquid) metals offer the possibility of much higher operating temperatures than can be obtained with water and have excellent heat transfer properties. Only one metal, mercury, is liquid at room temperature, and in any case it has far too high a neutron absorption cross section for use in thermal reactors. It also has relatively high vapor pressures, and the vapor is toxic. However, it is interesting to note that mercury has been used in power generation cycles, and several power stations in the United States were operated during the 1950s using mercury as a working fluid.

The only metals that combine the advantages of a relatively low melting point with low vapor pressure and low neutron absorption are sodium and potassium. Sodium and potassium are compatible with stainless steel at temperatures up to at least 800°C provided the liquid metal is kept free of oxygen. Sodium is the more abundant and cheaper to produce, and, furthermore, potassium can form compounds with oxygen that are explosive. In recent years, sodium has been the preferred liquid metal coolant, but in earlier reactors sodium-potassium mixtures (NaK) were frequently employed. The mixtures could be made liquid at room temperature and the pipework did not require heating in order to keep the coolant molten during periods of shutdown.

Sodium has been the primary choice as the coolant for fast breeder reactors. Referring to Table 3.1, we see that it has a higher thermal conductivity, though a lower specific heat, than water. For a given heat removal, the flow rate required is five times higher for sodium than for water. However, the overriding advantage of sodium is its high boiling point, which allows sodium-cooled reactors to operate near atmospheric pressure while maintaining a wide difference between the operating temperature and the boiling point. The operating temperatures are sufficiently high for the sodium stream to evaporate water at high pressure to produce a high thermodynamic efficiency. However, there are a number of problems in using sodium as a coolant.

1. Because sodium is very reactive toward oxygen and water, contact must be avoided. An inactive cover gas such as argon is needed above all the sodium levels in the reactor system. This gas must be kept free of oxygen contamination. In the steam-generating system, the sodium heats tubes that contain the high-pressure evaporating water. These evaporators are one of the main sources of trouble in sodium-cooled reactors. Very small leaks can be tolerated, but they cause contamination problems due to the formation of sodium hydroxide, which is corrosive. Large leaks can cause explosive interactions between the sodium and water, giving rise to hydrogen generation, and the hydrogen itself is explosive. Contamination of the reactor vessel can be minimized by using an intermediate heat exchanger (Chapter 2); alternatively, some designs use double-walled tubes to enhance the separation between the two fluids in the steam generator. In the design of sodium-cooled reactors, provision is made to accommodate such sodium-water reactions safely. For instance, the steam generator can be isolated and its sodium content ejected through a special system that allows the hydrogen generated to be discharged through an outlet stack.

2. The primary coolant circuit becomes very radioactive through the formation of sodium-24, which has a 15-h half-life; the existence of this isotope in the primary coolant is another reason for having an intermediate exchanger between the coolant and the steam generator. In practice, in view of the fairly short half-life, this isotope creates no particular difficulty associated with maintaining the reactor circuit, though sufficient time must be allowed for it to decay to a low level before the circuit is worked on.

To someone who has witnessed school chemistry laboratory experiments in which small pieces of sodium are dropped into water and has observed the dramatic and explosive effects, the prospect of using this metal as a reactor coolant must seem rather horrifying. However, when contained within a reactor circuit, and with proper precautions taken to deal with any potential effects of its contact with the steam generator water, sodium is a surprisingly benign and extremely efficient coolant.

3.6 BOILING COOLANTS

There are a number of advantages in cooling a reactor core with a coolant that vaporizes (boils) in the core itself.

1. The vapor produced can be fed directly to a turbine, and power can be generated without an intermediate heat exchanger and/or vapor generator.
2. Boiling coolants are very efficient in heat transfer (see Section 3.3).
3. The evaporation process in the reactor core produces a mixture of vapor and liquid, which has a much lower neutron absorption than a liquid and at the same time maintains a very high heat transfer efficiency. As the proportion by volume of vapor in the coolant (commonly called the *void fraction*) increases, the neutron absorption decreases and there is an increase in the reactor neutron population, or the *reactivity*. If the coolant also acts as a moderator, the neutron population will decrease. Thus, reactors with boiling coolants that also serve as the moderator commonly have a decrease in neutron population with increasing void fraction, or a *negative void coefficient*. If the demand for steam from the reactor increases, therefore, the natural tendency of the reactor is to start to shut itself down, and the control system must be designed to accommodate this effect. In reactors of the pressure-tube type with separate moderators (e.g., graphite), there can be a *positive void coefficient* and the reactivity increases unless action is taken to offset the effect. It is noteworthy that when sodium boils in a fast reactor, where there is no moderator, an increase in reactivity is observed since there is a positive void coefficient in this case also.

The main disadvantages of boiling coolants are as follows:

1. The highly efficient boiling process can degenerate into an inefficient, essentially vapor-cooling process rather abruptly due to the phenomenon of dry-out or burnout, as described in Section 3.3.
2. Using vapor generated directly in the reactor core in the power generation system means that the latter system is somewhat radioactive, requires special design, and has increased maintenance and operating costs.
3. The rather complex behavior associated with the void coefficients, as described above, can also be a disadvantage.

Liquid-cooled reactors can inadvertently become boiling-liquid-cooled reactors in the event of a power excursion or a loss-of-coolant accident. We shall discuss this in detail in Chapter 4.

3.6.1 Water

Water is the most commonly used boiling coolant, and about 30% of the world's nuclear reactors are boiling-water reactors (BWRs). These reactors were described in Section 2.4.

Many of the features of water as a boiling coolant are identical to those of water as a liquid coolant, which were described in Section 3.5. It should be noted that BWRs operate at much lower pressures than PWRs (7 rather than 15.5 MPa, 1000 rather than 2300 psia).

Using water as a boiling rather than a liquid coolant entails the additional important problem of radiolysis, whereby the water is decomposed into its constituent elements, hydrogen and oxygen, which are released into the vapor during the boiling process. The rate of recombination of the hydrogen and oxygen is much slower than in a system operated purely in the liquid phase, leading to higher concentrations of oxygen in the circuit fluid. Since the circuit is under stress due to the high pressure, a form of corrosion called *stress corrosion cracking* can occur, and this has presented a major difficulty in the operation of BWRs. It can be overcome by using more resistant materials, but replacing pipework in existing reactors is obviously an expensive process.

3.6.2 Liquid Metals

Boiling potassium coolants have been investigated for both terrestrial and space power systems, and a prototype space reactor employing boiling potassium has been operated. This coolant is attractive because it gives a very high thermodynamic efficiency, corresponding closely to the special case discussed in Section 3.2, with a high latent heat and a low specific heat. Thermodynamic efficiencies of the order of 55% (compared to 35–40% in water systems) are possible with a combined potassium and steam cycle. However, this form of coolant has not been seriously pursued, mainly because of the exotic materials required for the construction of the fuel cladding and the turbine.

3.7 ALTERNATIVE FORMS OF REACTOR COOLANT CIRCUITS

Since the first air-cooled nuclear reactor built under the squash court of the University of Chicago in December 1942, an amazing variety of nuclear reactors have been devised and many of them have been built. In all cases a coolant circuit was included; the main components of such circuits and the circuits applied in the most commonly used nuclear power reactors are described in Chapter 2. Of course, all reactor cooling circuits must include the reactor core itself, a means of circulating the coolant through the core, and a means of extracting the heat from the coolant in order to maintain continuous cooling of the reactor and at the same time (in power reactors) generate useful power. In power reactors the means of extracting the heat from the coolant is almost universally a heat exchanger, which produces high-pressure steam that can be used in a steam turbine to generate power. It is convenient to divide the various types of reactor circuits into three groups:

1. *Loop-type circuits.* The core itself is contained within a reactor vessel, and the primary coolant circulator and the steam generator are coupled to the reactor vessel by suitable pipe systems.

2. *Integral-type circuits.* The core, primary coolant circulator, and steam generator are contained within a single vessel, feedwater is fed to this vessel, and steam is taken from it to the turbine.

3. *Pool-type circuits.* The core and the primary coolant circulators are immersed in a pool of coolant. This arrangement is feasible only for unpressurized coolants such as sodium. The steam generator is usually outside the reactor containment vessel. This type of circuit is intermediate between the loop-type and integral-type circuits.

3.7.1 Loop-Type Circuits

The prime examples of this type of circuit are those used in the Magnox, pressurized-water, and CANDU reactors. The circuits for these reactors are illustrated in Figures 2.4, 2.8, and 2.11). In normal reactor operation there is a multiplicity of loops, as illustrated in Figures 3.5 and 3.6, which show the positions of the individual loops in the PWR and CANDU systems, respectively. Note that in the PWR the loops come together in the reactor core, whereas in the CANDU reactor they are always totally separate. This has important implications for safety considerations with these reactors, as we shall see in Chapter 4. A typical modern large PWR has three or four loops, depending on the size, each loop handling typically 300 MW of electric power production (corresponding to generation in the reactor core of 900 MW of thermal energy for each loop). Smaller reactors have two loops, with the size of the steam generators and other components within a loop kept approximately the same. Some

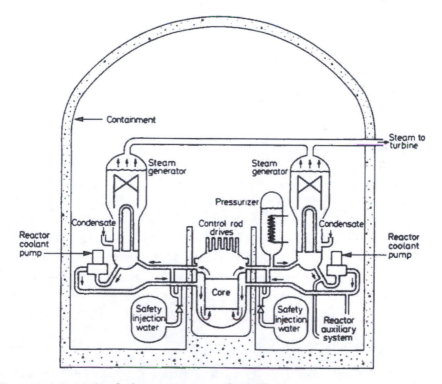

Figure 3.5: Example of a loop-type circuit: the PWR.

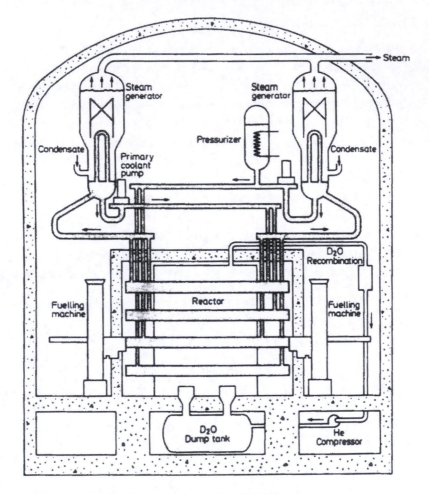

Figure 3.6: Example of a loop-type circuit: the CANDU reactor.

of the earlier PWRs had a four-loop design, notwithstanding their smaller overall size and much smaller output per loop (e.g., the Shippingport and Yankee Rowe reactors). The move toward standardization in the mid-1960s led to much larger reactors and much larger powers per loop.

3.7.2 Integral-Type Circuits

Typical examples of integral circuits are the advanced gas-cooled reactors and the type of reactor used most commonly for ship propulsion. The original CO_2-

cooled reactors (Magnox) were of the loop type. However, the development of large concrete pressure vessel technology allowed the incorporation of the steam generators and circulators inside the pressure vessel, as illustrated in Figure 2.5. A typical marine reactor is illustrated in Figure 3.7, where the circulators, steam generators, and reactor core are all encapsulated in a single steel pressure vessel.

The great advantage of the integral type of circuit is that all the primary circulating fluid is contained within the vessel, removing the need to circulate the primary fluid through connecting pipework to the steam generator. A possible accident source in the loop-type circuit is the rupture of one of the primary coolant pipes, and this is obviated in the case of the integral circuit.

3.7.3 Pool-Type Circuits

Perhaps a majority of nuclear reactors used for research are of the pool type, often referred to as "swimming pool" reactors. The core is immersed in a pool of

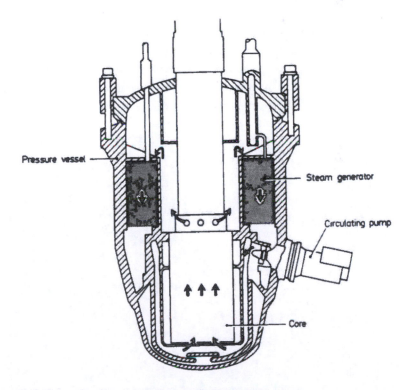

Figure 3.7: Example of an integral-type circuit: marine reactor.

light or heavy water, and heat exchangers are placed outside the reactor vessel to extract the heat. This principle of reactor design can generally be applied only when the primary coolant is an unpressurized liquid, which is the case only for the liquid metal–cooled fast reactor (see Figure 2.15 for an illustration of the circuit). Examples of reactors with this type of coolant circuit are the British prototype fast reactor (PFR) and the French Phenix reactor. However, it is also possible to design sodium-cooled fast reactors with loop-type circuits. Examples of such reactors are the Japanese sodium-cooled fast reactors JOYO and MONJU.

In the latter designs, the primary circuit pump and the intermediate heat exchangers are external to the vessel containing the reactor core, as illustrated in Figure 3.8. Thus, in these designs the sodium is pumped from the reactor vessel through pipes connecting it to the heat exchanger.

The advantages of the pool-type design are that there are no external pipes, which reduces the risk of pipe ruptures, and there are no connections to the tank containing the coolant pool below the liquid level, as illustrated in Figure 3.9. Moreover, in pool-type reactors, the large quantity of sodium in contact with the core can act as a heat sink in case of circulation failure. In fact, with a well-designed sodium-cooled fast reactor of this type, it is possible to ensure decay heat removal by natural circulation alone, and we shall return to this point in Chapter 4. The pool design, however, has the disadvantage that the main core structures are submerged under many thousands of tons of active sodium and are difficult to get at (to monitor their structural integrity) and to

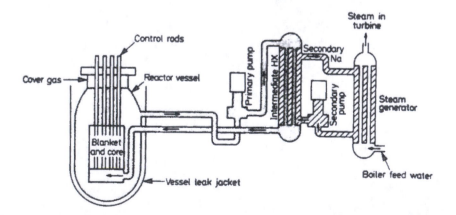

Figure 3.8: Example of a loop-type circuit: the liquid metal–cooled fast reactor.

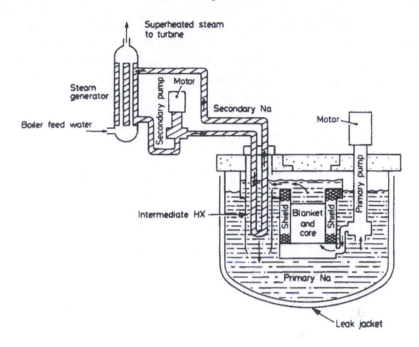

Figure 3.9: Example of a pool-type circuit: liquid metal–cooled fast reactor.

maintain. Access and maintenance are much easier in the loop-type reactor, but the existence of external pipework introduces the possible hazards of a loss-of-coolant accident.

3.7.4   Future Developments

Future development of nuclear reactors is aimed at improved performance—from both an economic and a safety viewpoint. Electricity utilities in both Europe and the United States are collectively defining these requirements in detail. A key issue in the application of future designs is their ability to be licensed in those countries wishing to deploy the design, in the same way as aircraft designs achieve their airworthiness certificates to operate internationally.

Most attention has been directed at advanced light-water reactors (ALWR). Two different approaches are being pursued to meet these improved-performance goals; they relate to evolutionary and passive designs, respectively.

Evolutionary designs are extensions of existing PWR and BWR plants building upon past experiences and using proven components but with enhanced safety features designed to reduce the probability of accidents and to mitigate their consequences. Specific examples of these plants and their vendors are the European pressurized water reactor—EPR (NPI); System 80 plus(™)/BWR 90 (ABB); ABWR (GE); and Sizewell B/APWR (Westinghouse/Mitsubishi).

Passive designs make effective use of natural physical processes such as gravity (control rod insertion), natural circulation/convection (to remove heat), evaporation-condensation, transient heat conduction (to provide heat sinks), stored energy in pressurized accumulators (to inject cooling water), and negative reactivity effects (to stabilize the chain reaction). Although there are drawbacks, the maximum use of such natural phenomena can simplify the design and reduce dependence on operator action. Specific examples of passive plants and their vendors are the simplified BWR (GE); PIUS (ABB); and AP600 (Westinghouse).

REFERENCE

Etherington, H. (1958). *Nuclear Engineering Handbook*, sec. 9.3, 9–91. McGraw-Hill, New York.

EXAMPLES AND PROBLEMS

1 Nuclear fuel center temperature

Example: Derive an expression for the temperature at the center of a nuclear fuel pellet assuming that the internal energy generation is uniform and the thermal conductivity is independent of temperature. A solid UO_2 pellet has a linear rating of 45 kW/m and a surface temperature of 600°C. The thermal conductivity of UO_2 is 2.7 W/m K. What is the center temperature of the fuel pellet?

Solution:

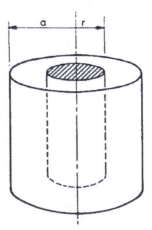

Suppose the rating of the fuel pellet, i.e., the total energy supplied as heat per meter of fuel, is R (W/m). Then the rate of energy release within a radius r is

$$\left(\frac{r^2}{a^2}\right)R$$

since the power produced is in proportion to the volume of fuel.

At equilibrium this rate of energy is conducted away from the cylindrical surface at r, i.e.,

$$-k\,2\pi r\left(\frac{dT}{dr}\right)$$

where k is the *thermal conductivity* and T is the temperature. So

$$-k\,dT = \left(\frac{R}{2\pi a^2}\right)r\,dr$$

and the temperature at radius r, $T(r)$, is

$$T(r) = T_o + \left(\frac{R}{4\pi k}\right)\left(1 - \frac{r^2}{a^2}\right)$$

and the center temperature is

$$T_{MAX} = T_o + \frac{R}{4\pi k}$$

where T_o is the temperature of the outside of the pellet ($r = a$). Note that the difference in temperature ($T_{MAX} - T_o$), when the energy release rate is expressed as a linear rating, is *independent* of the diameter of the pellet.

$$T_{MAX} = 600 + \frac{45}{4 \times \pi \times 2.7} = 1926°C$$

Problem: For the UO$_2$ pellet described in the example, calculate the maximum linear rating that would be possible if the center temperature were limited to 1500°C.

2 Figure of merit for a reactor coolant

Example: A figure of merit for a reactor coolant is given as

$$F = \frac{C_p^{2.8}\,\varrho^2}{\mu^{0.2}}$$

Derive this expression from a consideration of the ratio of pumping power P to heat output Q for a constant coolant temperature rise ΔT.

Solution: The pressure drop Δp across a channel of diameter D and length L is

$$\Delta p = \frac{1}{\varrho}\left(\frac{W}{A}\right)^2 \frac{2fL}{D}$$

where W is the flow rate of the coolant, A is the flow cross-sectional area, f is the friction factor, which for turbulent flow is proportional to

$$\left(\frac{WD}{A\mu}\right)^{-0.2}$$

and ϱ is the coolant density.

The pumping power $P\,(=\Delta p W/\varrho)$ is

$$P = \frac{1}{\varrho^2}\left(\frac{W^3}{A^2}\right)\frac{2fL}{D}$$

The heat output Q can be given in terms of the flow rate, specific heat, and temperature rise of the coolant:

$$Q = W C_p \Delta T$$

If we use this equation to eliminate W from the pumping power equation, then

$$P = \frac{1}{\varrho^2 A^2}\left(\frac{Q^3}{C_p^3 \Delta T^3}\right)\frac{2fL}{D}$$

but the friction factor f is proportional to

$$\left(\frac{WD}{A\mu}\right)^{-0.2} \text{ or } \left(\frac{QD}{AC_p \Delta T \mu}\right)^{-0.2}$$

Thus P is proportional to

$$\frac{2L}{A^2 D^{1.2}}\left(\frac{Q^{2.8}}{\Delta T^{2.8}}\right)\frac{\mu^{0.2}}{C_p^{2.8}\varrho^2}$$

Therefore, for given channel dimensions L, A, and D, heat output Q, and temperature rise of coolant ΔT, the pumping power will be a minimum when

$$\frac{\mu^{0.8}}{C_p^{2.8}\varrho^2}$$

is a minimum or the reciprocal is a maximum.

Problem: A new organic coolant is being considered for reactor cooling. At the condition obtained in the proposed reactor, its density is 862 kg/m³, its viscosity is 1.5 x 10⁻⁴ Ns/m³, and its specific heat is 2710 J/kg K. Calculate the figure of merit for this new coolant and compare the value obtained with those for other coolants given in Table 3.1.

3 Circuit designs for nuclear reactors

Problem: The development of nuclear power reactors has been such that for gas-cooled reactors the integral-type design has become standard (using a pressurized concrete vessel), whereas for water-cooled reactors the loop-type design is favored. Discuss the relative advantages and disadvantages of these alternative designs and establish why each is preferred.

BIBLIOGRAPHY

Bergles, A.E., Collier, J.G., Delhaye, J.M., Hewitt, G.F., and F. Mayinger (1981). *Two-Phase Flow and Heat Transfer in the Power and Process Industries*. Hemisphire, Washington, O.C., 719 pp.

Berglund, R.C. (1993). "Progress Report on GE Advanced Reactor Family." *Nuclear Europe Worldscan* 13 (March–April): 48.

Caso, C.L. (1993). "Advanced Designs for World Applications." *Nuclear Europe Worldscan* 13 (March–April): 50.

Catron, J. (1989). "New Interest in Passive Reactor Designs." *EPRI Journal* 14 (April–May): 4–13.

Collier, J.G., and J.R. Thome (1994). *Convective Boiling and Condensation*, 3d ed. Clarendon, Oxford, 596 pp.

Delhaye, J.M., and M. Giot (1981). *Thermohydraulics of Two-Phase Systems for Industrial Design and Nuclear Engineering*. Hemisphere, Washington, D.C., 540 pp.

French, H., ed. (1981). *Heat Transfer and Fluid Flow in Nuclear Systems*. Pergamon, Elmsford, N.Y., 582 pp.

Golay, M.W., and N.E. Todreas (1990). "Advanced Light Water Reactors." *Sci. Am.* 262 (April): 82–89.

Hall, W.B. (1958). *Reactor Heat Transfer*. Temple University Press, Philedelphia.

Hüttl, A., and J.C. Leny (1993). "Framatome-Siemens Co-operation on the European Pressurized Water Reactor (EPR)." *Nuclear Europe Worldscan* 13 (March–April): 43–45.

Kutateladze, S.S., and V.M. Borishanskii (1959). *Liquid-Metal Heat Transfer Media*. Consultants Bureau, 150 pp.

Lahey, R.T., and R.J. Moody (1977). *The Thermal Hydraulics of a Boiling Water Reactor*. American Nuclear Society, LaGrange Park, Ill.

Merilo, M., ed. (1983). *Thermal-Hydraulics of Nuclear Reactors*, 2 vols. International Topical Meeting on Nuclear Reactor Thermal-Hydraulics, Santa Barbara, Calif,. January 11–14, 1983, 1529 pp.

"The New Reactors" (1992). *Nuclear News* 35 (September): 66–90.

Runermark, J., and I. Tiren (1993). "ABB's Program for Evolutionary LWR's." *Nuclear Europe Worldscan* 13 (March–April): 46–47.

Tong, L.S., and J. Weisman (1979). *Thermal Analysis of Pressurized Water Reactors*. American Nuclear Society, LaGrange Park, Ill.

Winterton, R.H.S. (1981). *The Thermal Design of Nuclear Reactors*. Pergamon, Elmsford, N.Y.

4

Loss of Cooling

4.1 INTRODUCTION

A modern large nuclear power plant is a very complex piece of engineering with a wide diversity of components. In the design of such plants, careful consideration must be given to the effect of breakdowns of these components. In this chapter we shall be primarily concerned with those component breakdowns, or combinations of component breakdowns, that can give rise to an interruption in normal cooling. When such an interruption occurs, the fission reaction is rapidly terminated, but as we saw earlier (Section 2.2, particularly Table 2.2), heat generation continues after shutdown of the fission reaction due to the continuing decay of the fission products that have been generated. All reactor systems are provided with alternative means of cooling in order to remove this fission product decay heat in the event that the normal cooling system fails to operate. In Chapter 6 we shall consider the consequences of the alternative cooling system itself failing to operate, although this is a very remote possibility.

The design of a nuclear power station must encompass a number of *operational states* that can occur during normal operation of the reactor or as a result of some kind of fault. These operational states may be classified as described below and are summarized in Table 4.1.

1. *Normal operation and operational transients.* In addition to the normal operational state, as described, for instance, in Chapter 2 for the various reactor systems, the designer must think about *transients* that occur during operation. The term transient implies a nonsteady state of operation encountered in proceeding normally from one steady operating state to another. An example would be bringing the reactor from a "cold" condition up to full-power operation. This kind of operational transient must be taken into account in the design and the methods for achieving it worked out in the operating instructions for the reactor. For instance, to avoid damaging the structure of the reactor, there may be limits to the rate at which the temperature of the structure can be increased or decreased. To ensure economic opera-

tion of the reactor, many components that might require frequent maintenance are duplicated, and the rules for operation of the reactor must be carefully worked out to ensure that safe operation can be maintained even when some of the components are out of service. In order to operate the reactor economically, consideration must be given not only to the steady state but also to all the things that are likely to happen as a matter of course in the operation of a complex engineering plant.

2. *Upset conditions.* The word *upset* is used to describe all the kinds of faults that are not expected during operation but that can be reasonably expected to occur during the lifetime of a plant as a result of a variety of external events. Consider, for example, the case of lightning striking the power lines leaving the plant. A plant generating 1000 MW of electricity suddenly has no means of exporting this electricity to the grid. When electricity is no longer taken from the generator attached to the steam turbine, the turbine will increase in speed unless rapid action is taken to prevent such an occurrence. This action is to stop the flow of steam to the turbine and divert it directly into the condenser. The steam flow is reduced as rapidly as possible by using the control rods to stop the fission reaction—"tripping" the reactor. A turbine trip of this kind might be expected to occur for one reason or another about once every year, and it is important to design properly to accommodate it.

An interesting consequence of such a trip is that the power station, instead of being an exporter of electricity, immediately becomes an importer of electricity in order to drive the coolant pumps, instrumentation, and emergency cooling systems for the reactor. If the external power line has been broken, it is likely that no electricity can reach the site. Since the reactor has been tripped, it is no longer generating electricity and emergency power generation systems must be provided. These are usually diesel-driven generators, and normally several of them are installed in case one is being serviced or

Table 4.1 • Classification of Reactor Operating States and Frequency of Occurrence

Operating states for which the system is designed to cope:	
Normal operation	Continuous (apart from shutdowns for maintenance)
Operational transients	~10 per reactor year
Upsets	~1 per reactor year
Emergencies	1 in 100 reactor years
Limiting fault conditions (including design basis accident, DBA)	1 in 10,000 reactor years
Unprotected or beyond design basis accidents	1 in 1 million reactor years

fails to operate. This example of an upset transient is one of many that must be accounted for in design; others include loss of cooling water to the condenser due to the failure of a cooling-water pump, loss of feedwater to the steam generator, and reactor coolant pump trips.

3. *Emergency events.* Although operational transients are certain to occur and upsets are practically certain to occur during the lifetime of a plant, a number of events can be postulated that might have, say, a 1-in-10 chance of occurring in the lifetime of a particular plant. If we consider a sample of 10 plants, it is practically certain that one of these events would occur within 1 of the plants during its lifetime. A large country such as the United States has more than 100 reactors in operation; therefore, emergency conditions are likely to occur within one of the plants every few years. The reactor design must cope with such emergencies, although some damage to plant components may be expected as a result of the incidents. An emergency event would occur, for instance, as a result of breaks in small pipes in the reactor circuits, relief valves being stuck open, or fires within the plant electrical systems.

4. *Limiting fault condition.* It is possible to conceive of events, such as an earthquake, the complete severance of a main inlet pipe, or the complete severance of a steam line from the steam generator to the turbine, that would represent a severe accident to a reactor. Even though some accidents might occur only once in 10,000 years of reactor operation (though with 100 reactors operating, such an event might occur once every 100 years), reactors must be designed to meet these so-called limiting fault conditions safely. Although an *emergency event* (as described above) would not give rise to any release of activity off the reactor site, a *limiting fault condition* could give rise to extensive failure of the fuel canning and some consequent release of radioactivity off the site. The regulations set down by the national licensing bodies limit this release of radioactivity to a level that would not represent any significant risk to the public.

The reactor must be designed to meet the above operating states. Certain faults—for example, those related to coolant circulation pumps or gas circulators—can give rise to an interruption in normal cooling. When such an interruption occurs, the reactor is shut down by its automatic safety systems. But as we saw earlier, heat generation continues *after* shutdown of the fission reaction due to the continuing decay of the fission products that have been generated: the *decay heat.* So all reactor systems are provided with alternative means of cooling in order to remove this decay heat in the event that the normal cooling system fails.

The two most important safety systems are those associated with stopping ("tripping") the fission reaction within the reactor (the control rods) and those associated with providing an alternative cooling system, the so-called emergency core cooling system (ECCS). These *engineered safety systems* need to be brought into operation reliably when required.

This is done as a result of instrumentation signals received from sensors located around the plant that indicate when an unsatisfactory condition is being approached. They then initiate the action of the safety systems. This total reactor protection system has to be highly reliable. Such reliability is achieved through:

1. *Duplication.* Several sensors are used to measure critical parameters and several signal processors used to evaluate the signals. If all the sensors and processors are working, some of them are *redundant.* In a typical protection system, there are four identical sensors whose readings are compared. If two sensors of the four give identical signals requiring the activation of the safety system, action is taken. This allows for failure of two of the four systems.
2. *Diversity.* Different systems parameters are monitored to provide an indication of the same form of fault condition. Thus, two completely different signals, e.g., pressure and temperature, can be used to trip the reactor and/or initiate the emergency core cooling system for the same fault.

In some designs—for example, the British PWR, Sizewell B—the reactor protection system itself consists of two diverse systems: the primary protection system and the secondary protection system. The primary protection system is a microprocessor-based system that provides reactor trip and actuation of the engineered safety systems. The secondary protection system utilizes magnetic logic relays to initiate the reactor trip and engineered safety systems independent of the primary protection system.

As to the physical *trip* systems, advanced gas-cooled reactors have two separate systems for terminating the fission reaction: the first based on the control rods and the second on the injection of nitrogen, which is a neutron absorber, into the reactor gas.

Pressurized water reactors (PWRs) will shut down automatically if cooling water is lost from the core since this water is also the moderator. There are of course other systems. The normal one is based on control rods and the injection of boric acid, a neutron absorber, into the reactor cooling water. In addition, in some designs (Sizewell B) there is a completely separate second system for injecting very quickly large quantities of boric acid to deal with particularly severe faults.

Despite all these attempts to reduce the probability of failure of the protection system, it is difficult to demonstrate that such systems have a better reliability than 1 failure for every 10,000 times they are called into operation.

However, since the protection system itself is seldom called into operation (i.e., about once a year to meet an upset), the chance of failure is still remote.

It is now important to identify the vital supporting role of the operator. For faults of the type described above, no claims whatever are made on the operator for the detection of the fault and the safe shutdown of the reactor. This is achieved entirely automatically with high reliability. The role of the operator is really a management task of information gathering, planning, and decision making and only occasionally calls for more active control when routine operation is disrupted. Operators are highly trained—on simulators and on actual plant—and are regularly tested for competence.

It is a requirement in the design of the most recent British reactors that the operator should not need to intervene to control an abnormal condition for a period of at least 30 minutes after it begins. The automatic systems are designed to achieve this. During this period the operator needs essentially to monitor the proper functioning of the safety systems. He takes action only if the response of these systems is judged inadequate for some reason.

Operational transients, upsets, emergency events, and limiting fault conditions, as defined above, represent the range of conditions against which the plant is designed. The most serious of these conditions, the limiting fault condition, is often referred to as the *design basis accident* (DBA). It is possible to conceive of accidents that are more serious than the DBA and against which the reactor is relatively unprotected. Examples of such accidents are as follows:

1. Events that can be postulated but that are considered to be so unlikely that there is no justification for protecting the reactor against them. These might include the occurrence of a large earthquake in a zone where earthquakes do not normally occur and the direct crash of a large aircraft into the reactor with simultaneous destruction of the containment and the protection systems.

2. The occurrence of an upset, emergency condition, or limiting fault condition with the simultaneous failure of the protection system and/or the safety systems (for example, the emergency core cooling systems, ECCS). As indicated in Table 4.1, limiting fault conditions might occur once every 10,000 years. If the probability of failure of the emergency core cooling system was once in every thousand demands, a very severe accident leading possibly to the melting of the reactor core would occur once every 10 million years of reactor operation (i.e., 10^{-4} events per reactor year multiplied by 10^{-3} failures of the ECCS per event).

3. Although the designer tries to envisage all conceivable operational transients, upsets, emergency conditions, and limiting fault conditions, it is nevertheless possible that some event may happen that was not thought of. The most un-

predictable events are those that involve a sequence of multiple failures coupled with unanticipated responses from the reactor operators. It was this kind of sequence that occurred in the Three Mile Island accident, which we shall describe in more detail in Chapter 5.

Having discussed in this and previous chapters the basic principles of *controlling the nuclear reaction* and *cooling the fuel*, we need to introduce a third basic principle, that of *containing the radioactivity*. Collectively these three basic principles of reactor safety can be remembered as the Three Cs.

The term *containment* can be used to describe both a system for preventing the release of radioactivity to the general environment and the building in which a reactor is housed. Containment of radioactivity involves a multibarrier approach. What are these barriers?

1. Most radioactive fission products are retained where they are formed within the fuel, so the fuel matrix itself provides the *first barrier*.
2. The fuel is sealed in metal tubes—stainless steel for AGRs, zirconium alloy for water reactors. These are strong enough in all normal circumstances to contain all the fission products that escape from the fuel matrix. This is the *second barrier*.
3. The reactor pressure vessel that contains the core and the high-pressure coolant forms the *third barrier*.
4. And for many reactors there is the further barrier of the *containment building* itself—often a prestressed or reinforced-concrete-sealed and pressure-retaining building capable of withstanding external impacts and internal explosions.

The whole purpose of the safety systems provided on a reactor is to ensure that these separate barriers are not challenged and all remain intact. The safety limits are thus defined with this specific objective in mind.

In providing a framework for what follows in this chapter, it is useful at this stage to consider some basic principles related to the energy aspects of an accident. We may write the following simple *energy balance* for the reactor system:

$$\text{Energy in} - \text{energy out} = \text{energy stored}$$

As the reactor is brought up to power, some of the fission energy is stored in the reactor components as they are brought up to temperature. In particular, the fuel elements themselves store energy due to the large temperature gradient required to transfer the heat from them, as indicated in Figure 3.1. Once the reactor reaches steady state operation, energy is no longer stored and energy in equals energy out. Energy is also stored within the primary circuit coolant as a result of its heat capacity and, for a pressurized coolant, as a result of its high pressure. Any transient process causing a departure from steady state conditions will also cause a change in the stored energy. However, the above equation will continue to apply during the transient. Let us take two examples to illustrate this point:

1. If there is a failure of the secondary coolant passing, say, to a steam generator, then the output of energy from the system is reduced and the thermal energy storage of the system must increase, leading to an increase of temperature in all the primary circuit components. In some systems (such as pressurized-water reactors) this will also lead to an increase in the primary system pressure, and the consequences of this must be carefully evaluated.
2. If there is a failure in heat extraction from the fuel elements by the primary coolant, then the energy produced by the fuel elements must be stored within the fuel elements themselves, giving rise to a rapid increase in temperature.

The concept of the energy balances associated with transient conditions can also be applied to the case of heat release via breaks in the reactor circuit. These will lead to a loss of primary circuit coolant, reducing the amount (or *inventory*) of this coolant in the circuit. If the coolant is released from the circuit in the form of a vapor, it takes with it much more energy than if it is released in the form of a liquid, and this is advantageous in reducing the energy storage (e.g., the amount of heat stored in the fuel elements) during the transient situation. We shall return to this point in discussing the specific case of pressurized-water reactors (PWRs) in Section 4.3.

A very extreme case of heat retention in the fuel is that where no heat at all is removed from the fuel following a transient leading to a reactor trip. This case is illustrated in Figure 4.1. Initially, the fuel temperature becomes equalized, which gives rise to an initial relatively rapid increase in the fuel surface temperature as shown. The fuel element continues to heat up because of heat released during fission product decay, and this rate decreases with time as the fission products gradually disappear. The fuel will ultimately reach its melting point, however, and in designing for the various levels of transient condition it is obviously important to prevent this from occurring. The rate of rise of temperature of the fuel will depend on the initial heat rating, which determines the amount of fission products present at any given time. The temperature transients also differ from fuel to fuel and from reactor to reactor.

Having introduced the general principles of design in response to various transients, the types of occurrence and system design in response to them for

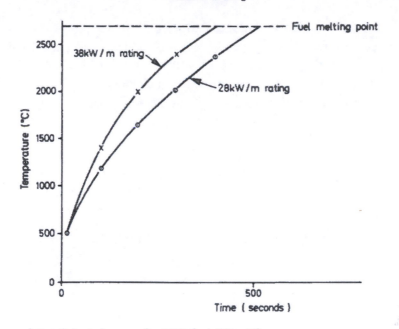

Figure 4.1: Adiabatic heat-up for PWR fuel (17 x 17).

water reactors (PWR, BWR, and CANDU), gas-cooled reactors, and fast reactors will be discussed. First, however, we will illustrate some of the points by using a homely example—making tea with a domestic electric kettle.

4.2 THE ELECTRIC KETTLE

The stages in tea making are illustrated in Figure 4.2. The electric kettle (referred to by our publisher as the "English samovar") is first filled with water (*a*) and then connected to the main electricity supply (*b*). Ultimately, the water in the kettle boils (*c*), and the tea is brewed (*d*).

We may represent the stages illustrated in Figure 4.2 in more scientific terms by plotting graphs (as shown in Figure 4.3) of the following quantities:

1. The amount of water present in the kettle—the inventory.
2. The power input to the element
3. The temperature of the water
4. The surface temperature of the element
5. The temperature of the element windings

As will be seen, the kettle has many of the characteristics of a reactor system. Since it does not have any recirculating coolant, its temperature rises until the boiling point of the liquid is reached, at which point boil-off of the liquid occurs. When the kettle is partially emptied (simulating a loss-of-coolant accident), the temperature of the surface of the heating element increases. The

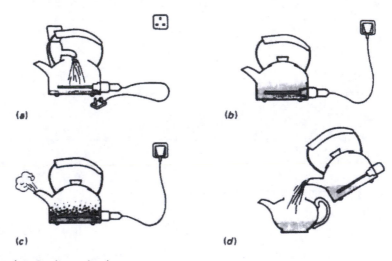

Figure 4.2: Cooling a kettle.

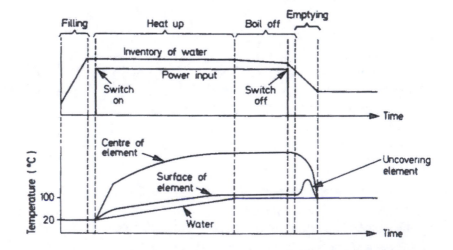

Figure 4.3: Analysis of kettle transient.

sequence of events shown in Figure 4.3 illustrates the case of rewetting or quenching, which occurs when the kettle is returned to its normal position and the remaining water quenches the hot heating elements with a very audible hiss. Furthermore, modern kettles are equipped with engineered safety systems; for instance, kettles often have a device that ejects the plug by means of a spring actuated by a bimetallic strip if the kettle boils completely dry during the boil-off period. In fact, even newer kettles have a device that detects the emission of vapor and switches the kettle off in a more easily reversible way when the water is boiled; this is another form of an engineered safety system.

The big difference between the kettle and a reactor system is that in the case of the kettle, the safety systems are able to switch off the power input completely and any overheating occurs principally as a result of stored energy within the heating element. In a reactor, however, fission product heating continues at a low rate relative to full power after the reactor has been shut down.

It is also interesting to compare the kettle with the reactor in the context of failure of the engineered safety systems. Plug ejection might occur several times during the lifetime of a kettle; reinstatement of the operation of the kettle can be achieved immediately afterward. However, should the plug fail to be ejected (due to, say, corrosion of the spring), then heating of the element will continue and can lead to its melting. Failure of the engineered safety systems leads to the need for a major repair of the kettle, and this would certainly also be the case in the nuclear reactor.

4.3 PRESSURIZED-WATER REACTOR

On a worldwide basis, the PWR is the most common power-generating reactor. It is appropriate, therefore, to deal with the various operating states and postulated accident conditions for this reactor in some detail, using the framework laid out in Section 4.1.

4.3.1 Operating States of the PWR

The situation in *normal* operation of a PWR is illustrated in Figure 4.4. The primary circuit consists of a pump that passes water at 292°C from the steam generator through the reactor core, where it is heated to 325°C (it does not boil at this temperature since it is at high pressure). This hot water passes back through the U-tubes in the steam generator, where it cools down to 292°C; the water on the secondary side of the steam generator is boiled to generate steam, which passes out of the containment to the turbines, is subsequently condensed, and returns through the feedwater pump to the secondary side of the steam generator. Also shown in Figure 4.4 are the various circuits for emergency core cooling water injection into the primary circuit (i.e., the emergency core cooling system). These consist of:

1. The accumulators. These are large vessels containing water that are pressurized with nitrogen gas. They are connected to the primary circuit via automatic valves, which open if the primary circuit pressure falls below a preset level (typically 40 bars).
2. A high-pressure injection system (HPIS). This allows pumping of water into the system at pressures of about 100 bars, though normally at a relatively low rate.
3. A low-pressure injection system (LPIS). This allows water to be pumped at a high flow rate into the reactor, provided the reactor is at a low enough pressure (typically below 30 bars).

The combination of emergency core cooling injection systems thus allows a response to a variety of reactor depressurization and loss-of-coolant accidents.

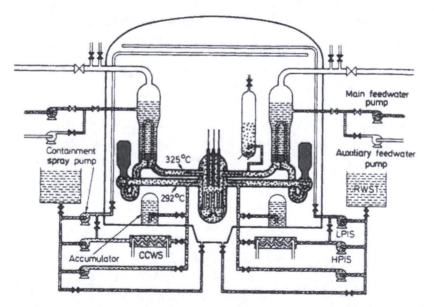

Figure 4.4: Diagrammatic representation of PWR primary and secondary circuits and the emergency cooling systems.

If water escapes from the primary circuit, it collects in a sump at the bottom of the containment vessel and may be recirculated from there through the ECCS pumps back into the primary circuit. In the LPIS, the flow passes back to the reactor through a heat exchanger, where it is cooled by the component cooling water system (CCWS). This provides a means of long-term decay heat removal from the reactor in the event of a loss-of-coolant accident. Note that the LPIS pumps can also be used to inject a spray of water into the containment to condense any steam present in the containment, thereby reducing the containment pressure in the event of an accident.

It is helpful in discussing PWR operational states to represent the operation in terms of a pressure/temperature map as illustrated in Figure 4.5. The solid line in Figure 4.5 represents the saturation temperature (or boiling point) as a function of pressure. The PWR must operate at a temperature to the left of this line to ensure that steam is not formed in the reactor. Figure 4.5 presents the operating conditions, showing the inlet and outlet temperatures at the operating pressure. The reactor pressure control is achieved in the pressurizer (see Figure 4.4) by having a body of liquid in contact with vapor at the saturation pressure. By raising or condensing steam within the pressurizer, the reactor circuit (which is connected to the pressurizer) is maintained at a fixed pressure. Thus, in terms of Figure 4.5, the pressurizer operates on the saturation curve as shown.

The reactor may reach the saturation condition by either increasing temperature or decreasing pressure. The most common way to reach saturation conditions is by depressurization, as illustrated. If the depressurization occurs by

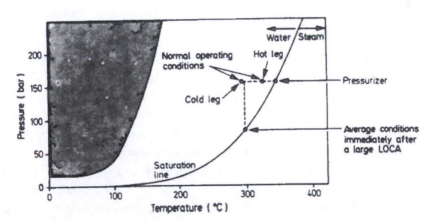

Figure 4.5: PWR operating conditions

means of a leak from the primary circuit, the initial rate of depressurization is extremely high. Once saturation conditions are reached, however, the rate of depressurization is much slower and may even reverse, with the reactor increasing in pressure for a short time.

Start-up and shutdown of the reactor must be carried out very carefully to avoid transients that would bring the reactor into a saturated state, with consequent vapor generation. It is also very important to avoid pressurizing the reactor vessel at too low a temperature. Doing this may cause existing small and insignificant defects in the vessel to extend and form significant cracks. The zone shown on the left-hand side in Figure 4.5 must also be avoided during operational transients such as start-up and shutdown. Thus, there is a "window" for operation that is bounded at both low temperature and high temperature as illustrated. In practice, the reactor is brought to its operating condition rather slowly over a period of about 24 h. A controlled return to cold shutdown also takes about 24 h.

The upset operating states of a PWR can be categorized as follows:

1. *Upsets leading to a change in the primary-side coolant inventory.* This could be (as illustrated in Figure 4.6) a loss of fluid through a relief valve or through some other service line to the reactor. The primary-side inventory may also be increased by inadvertent pumping of water into the circuit through the high-pressure charging pumps. In the latter case, the pressurizer may become totally flooded with water and pressure control may be lost.

2. *Upsets in the secondary-side heat removal capability.* This could include loss of feedwater supply or changes in feedwater temperature, maloperations of the main steam-isolating valves, a turbine trip, or maloperation of pressure-regulating valves and/or safety valves (see Figure 4.7).

3. *Other upset conditions* (see Figure 4.8). These include inadvertent maloperation of the control rod system and the possibility of a trip on one of the main reactor coolant pumps.

Emergency events in a PWR include (as illustrated in Figure 4.9) stuck-open pressure relief valves, a small break in the steam line, a small break in the primary circuit inlet pipe, and a loss of flow on *all* the reactor coolant pumps.

Limiting faults (defined in Section 4.1) in a PWR system are illustrated in Figure 4.10 and include a large break in the outlet steam line, a large break in the inlet primary circuit pipe, a steam generator tube rupture, the seizing up of the rotor on one of the main coolant circulating pumps, and the failure of a control rod mechanism housing (a control rod ejection accident). Of these, perhaps the most famous and most widely considered is the primary circuit inlet pipe failure (the design base accident for the PWR).

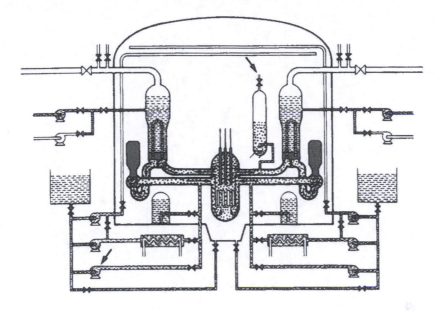

Figure 4.6: PWR upset conditions: control of primary-side inventory.

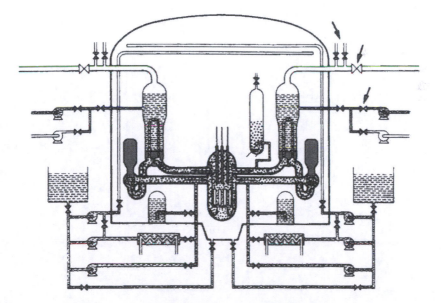

Figure 4.7: PWR upset conditions: control of secondary-side heat removal.

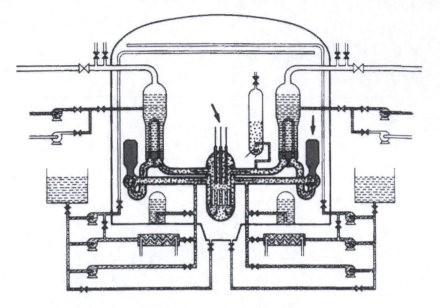

Figure 4.8: PWR upset conditions: other initiating situations.

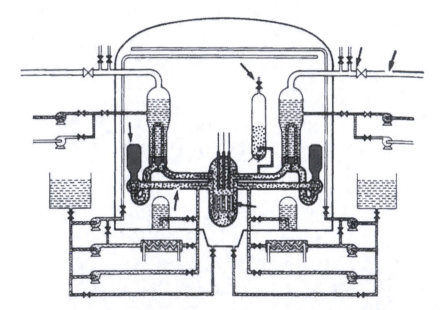

Figure 4.9: PWR emergency conditions.

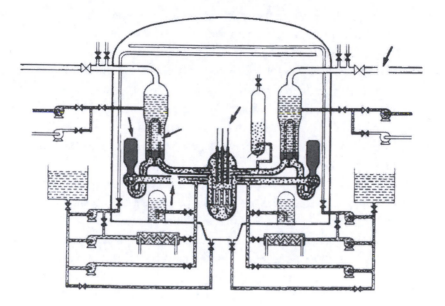

Figure 4.10: PWR limiting faults.

4.3.2   Energy Balances in the PWR under Fault Conditions

A typical PWR generating about 1100 MW(e) of electrical power would have a decay heat generation of about 200 MW(t) immediately after shutdown. This compares with 3400 MW(t) thermal energy generation immediately before shutdown. Removal of this decay heat is well within the capability of the low-pressure cooling system illustrated in Figure 4.4 *provided the reactor can be depressurized rapidly enough to bring these into operation.* Alternatively, if the steam generators can be operated effectively with the auxiliary feedwater system that is automatically switched on when the reactor trips, the decay heat can be removed via the steam generators, even at high pressure. A major difficulty arises when neither of these systems can be brought into play for reasons that will be described in Chapter 5. This is what happened in the Three Mile Island accident.

If the low-pressure cooling system and the steam generators are unavailable as a cooling mechanism, the only recourse is to feed water into the system via the high-pressure injection systems and the charging pumps (the pumps used to maintain the inventory of the system under normal operating conditions), the injected water bleeding out through the break. It is interesting to consider how the exiting fluid carries energy with it. The system is illustrated schematically in

Figure 4.11. If the water fed to the reactor is evaporated and exits as steam, this represents the maximum rate of energy release possible. If, on the other hand, the fluid leaves the reactor in the form of liquid water, not only is the discharge rate high (reducing the coolant inventory in the system) but the energy contained in the discharge is low relative to that in steam at the same mass flow rate. For these reasons, it is preferable to discharge steam rather than water. Discharges in the upper part of the circuit usually contain more energy than those in the lower part, where the existence of liquid water is more likely under transient accident conditions.

Taking the case of steam ejection from the reactor circuit, one can estimate that the maximum rate of ejection corresponds to the release of 17,000 MW of energy per square meter. To eject the steam that could be generated by the decay heat just after shutdown, a hole of area 0.011 m^2 would be required, corresponding to a hole diameter of 12 cm. The hole size required to reject the decay heat as a function of time from reactor shutdown (taking into account the decrease in decay heat rate as a function of time; see Table 2.2) is shown in Figure 4.12. One hour after shutdown the required hole size has dropped to 3.8 cm.

If the actual break size is bigger than that required to release the energy in the form of steam, the energy lost will be greater than that being generated and this will result in depressurization of the primary circuit. Such a depressurization may quickly lead to actuation of the low-pressure emergency heat removal systems. However, if the break size is smaller than that required to remove the energy, then energy will be stored within the reactor system, leading to

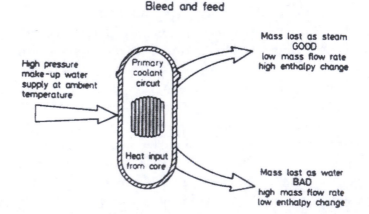

Figure 4.11: Energy outflows as steam and water.

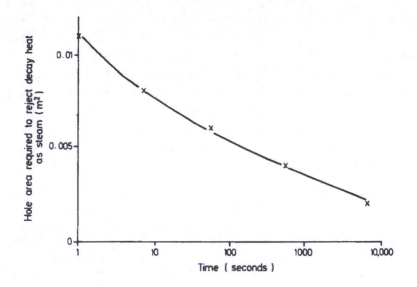

Figure 4.12: Hole sixe to remove decay heat as steam.

pressurization of the primary coolant. The system may be partly controllable if the power-operated relief valves (PORVs) can be opened to increase the escape of steam and facilitate energy release. The PORVs are located on top of the pressurizer, and a typical PWR would have two such valves with a total flow area of about 0.002 m², giving an energy release capacity as steam of about 34 MW. This is clearly much lower than the 200 MW of energy release corresponding to the decay heat immediately after reactor shutdown. In fact, it might be advisable to consider increasing the size and/or number of PORVs in future reactor designs to allow a higher rate of energy release.

If a break occurs and the available PORV area is insufficient to allow the energy release, the reactor system will continue to pressurize, ultimately actuating the spring-loaded safety valves, whose total area is likely to be sufficient to allow the energy release. However, the latter form of release is somewhat uncontrolled. The valves actuate and reseat at a specific pressure.

4.3.3   The Large–Break LOCA in the PWR

The classical design basis accident for a pressurized-water reactor is the large-break loss-of-coolant accident (LOCA). It is assumed that in this accident one of the inlet pipes from the circulating pump to the reactor vessel is completely broken and moved apart to allow free discharge of the primary coolant from both broken ends. This kind of break is called a "double-ended guillotine" break or a "200%" break. Because this break is commonly believed to represent about the worst accident that could happen to a water reactor circuit, it has been chosen as the basis for the design of the emergency response systems.

The sequence of events following the break is shown in Figures 4.13 to 4.17, which illustrate the situation in the whole reactor circuit. A more detailed illustration of the events within the reactor vessel itself is given in Figure 4.18. The main phases are as follows:

1. *Blowdown phase.* Under normal operation (Figures 4.4 and 4.18*a*), water flows through the inlet pipes (the *cold legs*) to the reactor vessel, down the annular space around the core, up through the core, and out through the vessel outlet pipes (*hot legs*) to the steam generator. When a large break occurs in one of the cold legs, the contents of the reactor vessel and primary loops are *blown down* through the break as illustrated in Figure 4.13 and 4.18*b*. After a very rapid initial depressurization, the pressure falls more slowly due to the creation of a two-phase steam-water mixture in the vessel and circuit, the mass flow of such a mixture through a break being much lower than that for a single-phase liquid. After about 10 s, the pressure has

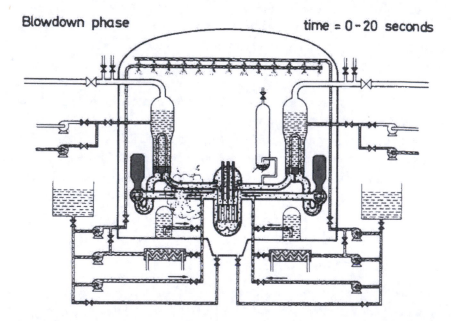

Figure 4.13: Large LOCA: blowdown phase (0–20 s).

By-pass phase   time = 20-30 seconds

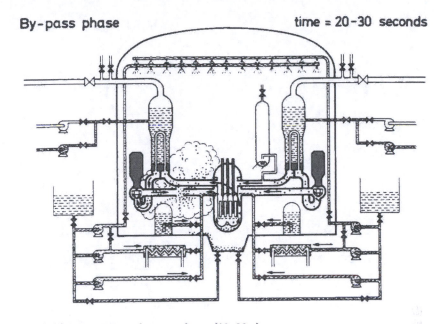

Figure 4.14: Large LOCA: bypass phase (20–30 s).

Refill phase   time = 30-40 seconds

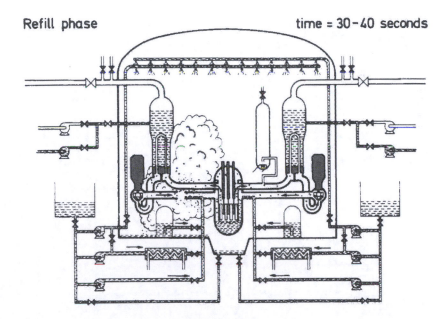

Figure 4.15: Large LOCA: refill phase (30–40 s).

Reflood phase   time = 40–250 seconds

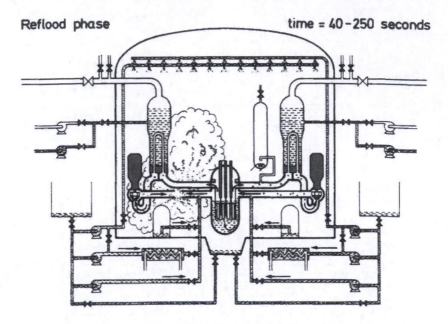

Figure 4.16: Large LOCA: reflood phase (40–250 s).

Long term cooling   time = >250 seconds

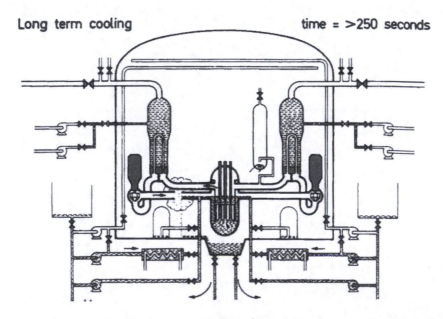

Figure 4.17: Large LOCA: long-term cooling phase.

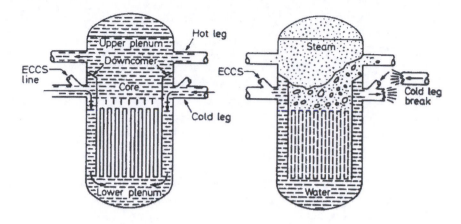

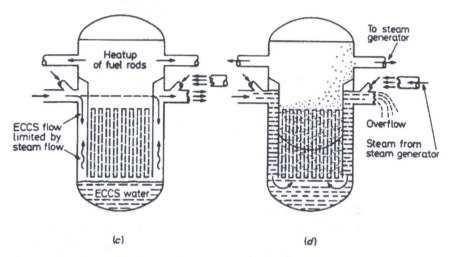

Figure 4.18: Events in the reactor pressure vessel during a large-break LOCA. (*a*) Normal operation; (*b*) blowdown phase; (*c*) refill phase; (*d*) reflood phase.

fallen to that for initiation of flow from the high-pressure injection system and the accumulators into the ECCS line in the cold legs.

2. *Bypass phase.* After the initiation of the ECCS, starting with the HPIS and the accumulators, there is still a significant *upward* flow of steam in the downcomer annulus through which the cooling water normally flows. This steam flow prevents the accumulator ECCS water from entering the region of the vessel below the core (the *lower plenum*), and the water simply bypasses the

upper part of the inlet annulus and flows out through the break, as illustrated in Figures 4.14 and 4.18c.

3. *Refill phase.* Filling of the lower plenum (the refill phase—Figures 4.15 and 4.18c) begins after further depressurization, when the steam flow up the annulus has dropped to a sufficiently low value that it can no longer restrict the ingress of ECCS water. By this stage, the LPIS system will also have been initiated. In a typical PWR, refilling of the lower plenum starts about 23 s after the initial break, and it takes 17 s to fill the lower plenum with liquid, ending this refill phase of the accident.

4. *Reflood phase.* At a very early stage in the blowdown phase, the core has dried out and the fuel element temperatures rise rapidly, after which they fall relatively slowly due to the existence of substantial steam flows in the core. Typically, the fuel element temperatures rise to around 1000°C. This leads to rupture of the fuel elements, which release gaseous fission products into the primary circuit and, via the break, into the containment vessel. The behavior of typical fuel elements as a function of temperature is shown in Table 4.2. When the lower plenum is filled, the reflood phase begins (Figures 4.16 and 4.18d), with the fuel elements being rewetted from the bottom upward. Essentially, a constant liquid head is maintained in the inlet annulus during this phase, with excess ECCS water overflowing through the break as illustrated. As the fuel elements rewet, a considerable volume of steam is formed and entrained liquid droplets flow before the rewetting front and pass into the upper plenum. The steam-droplet mixture passes from the upper plenum, through the steam generator, through the circulating pump, and back into the cold leg, flowing out through the break. The water droplets tend to evaporate in the steam generator due to the backflow of heat from the secondary-side (still hot) fluid. The resistance presented by the outflow route causes a back pressure in the upper plenum, which restricts the rate at which the reflood can take place. This phenomenon is often referred to as *steam binding.* The highest resistance of the upper plenum, through the steam generator and circulating pump, to the break would occur when all of the droplets issuing from the core passed to the steam generator and the circulating pump rotor was locked stationary. However, the resistance is much reduced, and the flooding rate greatly increased, if the droplets deposit out on the upper plenum structures and thus are not carried out of the vessel, and if the pump rotor is still rotating.

5. *Long-term cooling.* In the long term, the situation is as illustrated in Figure 4.17. Water is passed to the unbroken cold leg from the LPIS injection pump and maintains a head of liquid that drives water through the core by natural circulation. Steam maybe generated in the core and may escape with the overflow water through the break as illustrated. This generated steam is condensed by sprays in the containment, which are also fed from the LPIS pump.

Table 4.2 • Temperatures at Which Significant Phenomena Occur during Core Heat-Up

Temperature (°C)	Phenomenon
350	Approximate cladding temperature during power operation.
800–1500	Cladding is perforated or swells as a result of rod internal gas pressure in the postaccident environment; some fission gases are released; solid reactions between stainless steels and Zircaloy begin; clad swelling may block some flow channels.
1450–1500	Zircaloy steam reaction may produce energy in excess of decay heat; gas absorption embrittles Zircaloy, hydrogen formed. Steel alloy melts.
1550–1650	Zircaloy-steam reaction may be autocatalytic unless Zircaloy is quenched by immersion.
1900	Zircaloy melts, fission product release from UO_2 becomes increasingly significant above 2150 K.
2700	UO_2 and ZrO_2 melt.

In carrying out the design of a PWR, two types of calculations are usually employed, based on an evaluation model or on best-estimate methods. With an *evaluation model*, the various phenomena are represented by equations and assumptions that are postulated to give the worst conceivable result. For instance, it is normally assumed that there is no penetration of ECCS water during the blowdown phase. In *best-estimate methods*, the best available physical models are used for the various phenomena and an attempt is made to calculate the system behavior on the basis of these models. It should be pointed out, however, that the calculation of two-phase flows, particularly for the rapid transient conditions and large pipe sizes encountered in reactors, is still at an uncertain stage. As explained in Chapter 3, two-phase flows are very complex and in many respects poorly understood. It would be unsatisfactory to rely on two-phase modeling as a basis for reactor design. Some critics claim that the uncertainties in two-phase flow predictions imply that the reactor is unsafe. We do not share this view. From our long experience, we would agree with the assessment of the current state of modeling of two-phase flows but disagree that the reactor *design* is based on the results of such modeling. This design must satisfy very conservative criteria that do not depend on knowing about the details of two-phase flow behavior.

Figure 4.19 shows the variation of peak clad temperature as a function of time calculated from the evaluation and best-estimate models, respectively. The continuous line was calculated by using the conservative evaluation approach; the best-estimate values are shown as error bars.

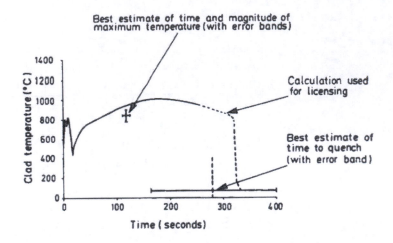

Figure 4.19: Variation of peak clad temperature with time for a large-break LOCA.

4.3.4 The Small–Break LOCA

Before the accident at Three Mile Island, most attention was focused on the postulated large-break LOCA. However, the Three Mile Island incident sharply focused attention on the fact that a small break (typically up to sizes where the reactor remains pressurized despite the break, say, up to 12-cm-diameter holes) in the primary circuit was, in fact, much more likely. At Three Mile Island this small break was due to a stuck-open power-operated relief valve. It could, however, also have occurred as a result of the break in one of the large number of small pipes attached to the primary circuit. Figure 4.20 shows a histogram of the number of pipes attached to the circuit as a function of pipe size and cross-sectional area. Also shown is the percentage area relative to the main coolant pipes. Figure 4.20 is for a German PWR design, but the result is likely to be much the same for other designs.

The most important difference between small-break and large-break accidents is that in the former the reactor depressurizes relatively slowly. Reactor pressure as a function of time after the break is shown for various break sizes in Figure 4.21. Since the core may remain at a high pressure in a small-break LOCA, it is not possible to activate the accumulator or low-pressure injection system until late in the accident.

A typical sequence for a small-break LOCA is illustrated in Figures 4.22 to

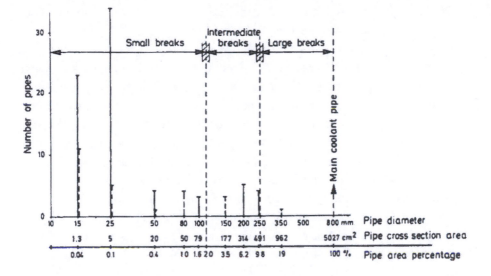

Figure 4.20: Connection pipe diameter/cross section/percentage spectrum of a PWR. Solid lines: primary loop system; dashed lines: pressurizer.

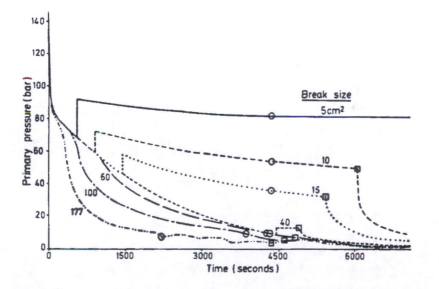

Figure 4.21: Primary pressure versus time for small-break LOCAs in PWR. (O) Primary temperature 175°C; (□) reflood tank empty. (Two HPI pumps; reflood tanks 4 x 286 m³; no LPI pumps.)

4.26. As with the large-break LOCA, the most serious effects are found when the break is in the reactor inlet pipe (the cold legs).

Following the initiation of the break, the pressure falls and the reactor trips. As the pressure falls below about 100 bars, the high-pressure injection system comes on. The pressure continues to fall to around 70 bars, when the hottest liquid in the circuit starts to vaporize and produce steam. First, the water in the pressurizer vaporizes. As the saturation condition is reached throughout the hotter parts of the primary circuit, steam bubbles form and because the pumps are stopped, will settle out in the upper part of the reactor as shown in Figure 4.22 There has been considerable controversy about whether to leave the circulating pumps operating or to stop them during a LOCA. If left on, they may assist in circulating liquid through the core, promoting its cooling. On the other hand, they may aid the loss of fluid by pumping it out through the breach. Current rules for operating PWRs indicate that the pumps should be stopped, this being considered to give the balance of advantage in general.

As a result of depressurization, steam forms and collects in the upper head of

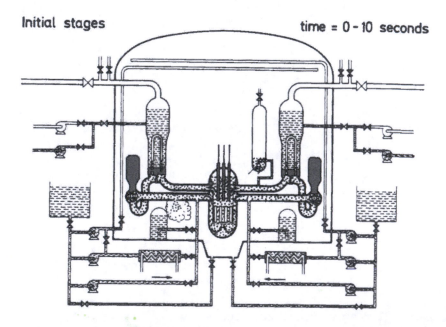

Figure 4.22: Small-break LOCA: initial phase.

the reactor but cannot escape via the breach because of the arrangement of the pipework. However, the loss of water is quite rapid, though it removes a relatively small amount of energy from the system. The water drains down to the level of the water inlet-outlet pipes on the reactor pressure vessel (the "nozzles") in about 250 s, during which the pressure may be maintained at a high level and still inhibit the actuation of the accumulators and LPIS.

During this phase, the steam generators are voided (i.e., their original water content is lost through the break) on the primary side, the steam from the reactor having access to the steam generators, as illustrated in Figure 4.23. Obviously, the steam generators represent a potential heat rejection source, with the steam from the reactor core being condensed in the steam generator tubes and flowing back down the tubes and into the core Figure 4.23). However, this beneficial action is conditional on the *secondary* side being at a sufficiently low pressure (and corresponding low saturation temperature) to allow heat to be extracted via the secondary circuit. Assume, for instance, that the secondary-side pressure remains at its normal operating value of 70 bars (1000 psia) while

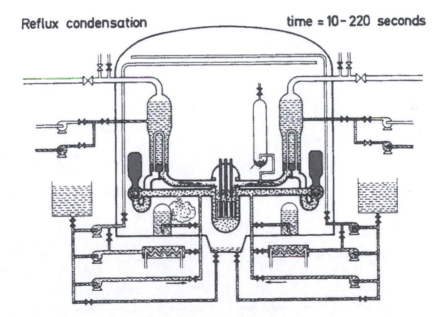

Figure 4.23: Small-break LOCA: reflux condensation.

the steam from the primary system is arriving at the steam generators also at 70 bars. In this situation the saturation temperatures are identical and no condensation can take place. Therefore, it is imperative in this type of accident that secondary-side cooling or depressurization is carried out.

Because of the continuing loss of water from the system, the core begins to dry out from the top downward (so-called core *uncovery*), as illustrated in Figure 4.24. The system is still pressurized due to the formation of steam, which cannot escape through the cold leg break, being blocked by the water in the vessel and the pump. The pump has a U-bend under it (the pump *loop seal*), and this remains full of water and blocks the steam from flowing from the vessel through the steam generator and pump to the break. The level in the pump side of the loop seal near the break (i.e., left-hand side of the loop seal at the left in Figure 4.25) is roughly equal to the level of water in the vessel since the two levels are connected by the (voided) steam generator and pump. Only when the levels reach the bottom of the U-bend can the steam pass the loop seal. At this point, steam from the core passes through the pump and out along the cold leg to the break. This results in a rapid depressurization. The water in

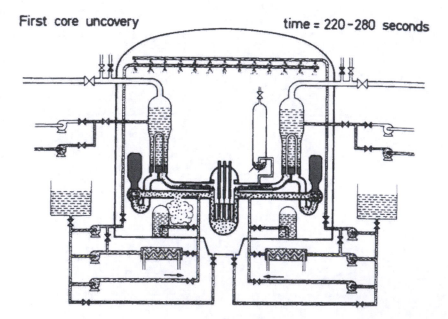

Figure 4.24: Small-break LOCA: first core uncovery.

Loop seal blowout
second uncovery

time = 280-310 seconds

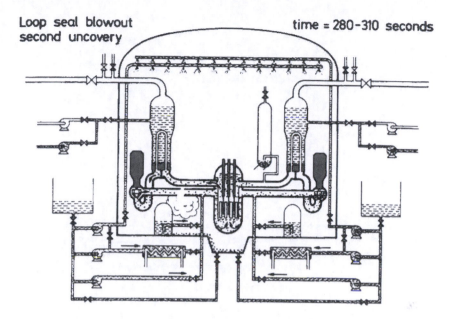

Figure 4.25: Small-break LOCA: loop seal blowout and second core uncovery.

the remaining part of the core partially vaporizes, and the mixture of water and steam bubbles formed rewets the upper part of the core. As depressurization proceeds, the core may be dried out again (as in a large-break LOCA). However, the depressurization permits actuation of the accumulator and LPIS systems, and these rapidly reflood the core and bring it to a cold condition. In the longer term (usually longer than 350 s) heat is extracted in the way illustrated for the large LOCA in Figure 4.17.

Figure 4.27 shows the variation of water level and fuel clad temperature for small-break LOCAs with two different equivalent break diameters. The level of the two-phase mixture in the vessel is shown. Normally, the cooling is good for regions of the core that are in contact with this mixture, the core being overheated above this level. The mixture level is much higher than the level of the liquid without steam bubbles would be. The latter level is referred to as the *collapsed liquid level*, and the phenomenon of increase of level due to the presence of the bubbles is termed *level swell*. A similar phenomenon occurs in dispensing glasses of beer.

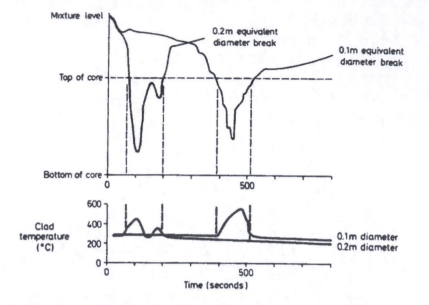

Figure 4.26: Small-break LOCA: mixture level and clad temperatures.

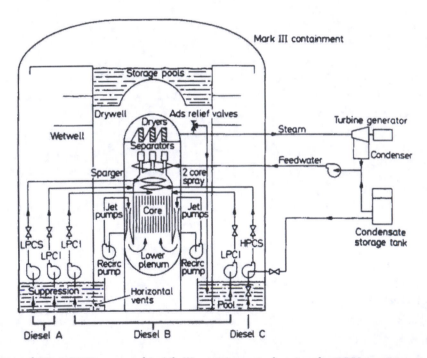

Figue 4.27: A BWR arranged in Mk III containment, showing the ECCS provisions.

4.3.5   Alternative ECCSs

The descriptions above apply mainly to a typical U.S. PWR, where the emergency core cooling water is injected into the coolant inlet pipes (the cold legs) only. Alternative ECCSs have been used, the most important being that used in the German PWR, where the emergency core cooling water is injected into both the cold legs and the hot legs (the coolant outlet pipes from the reactor vessel). Such combined injection is claimed to offer advantages in more rapid quenching of the core and in lower peak cladding temperatures during a large-break LOCA. In the case of the small-break LOCA, it is claimed that faster depressurization occurs, allowing early actuation of the accumulator and LPIS systems.

4.4   BOILING-WATER REACTOR

The boiling-water reactor, like the PWR, has multiple provisions for cooling the core in the event of an unplanned depressurization or loss of coolant within the reactor. A typical BWR emergency core cooling system is illustrated in Figure 4.27 It is composed of four separate subsystems, namely the high-pressure corespray (HPCS) system, the automatic depressurization system (ADS), the low-pressure corespray (LPCS) system, and the low-pressure coolant injection (LPCI) system.

The HPCS pump takes water from the condensate storage tank and/or the pressure suppression pool as shown in Figure 4.27 The water in the system is piped into the vessel and feeds semicircular perforated rings that are designed to spray water regularly over the core and onto the fuel assemblies. This system operates over the full range of reactor pressure and is activated when the water level in the reactor drops below a preset level or the pressure in the containment vessel reaches a high value.

If the HPCS cannot maintain the water level or if it fails to operate, the reactor pressure is reduced automatically by operation of the ADS, which discharges fluid from the vessel into the pressure suppression pool. The depressurization allows the LPCI and LPCS systems to come into operation, and these provide sufficient cooling. The LOCS pump takes its water from the suppression pool and discharges from a circular perforated pipe in the top of the reactor vessel above the core; it is actuated in much the same way as the HPCS. The LPCI system is used for residual heat removal on a long-term basis.

4.4.1   Large–Break LOCA in a BWR
(the Design Basis Accident)

The most serious accident considered for the design basis of a BWR begins with
the rupture of one of the pipes connecting the (external) circulating pump with
the reactor vessel as illustrated in Figure 4.28 This initial rupture produces a
more gradual depressurization than is the case in a PWR, since the pipe is considerably smaller than the main pipework in a PWR system (50 cm compared
with 80 cm in a PWR). Other factors restricting the depressurization rate are the
facts that the reactor vessel contains about 40% steam by volume and that the
steam line is shut off within a few seconds, isolating the vessel from the main
heat sink (the turbine) so that the system coolant can escape only from the
break. Although the flow in the damaged loop would reverse due to the break,
core cooling is maintained during the early part of the accident since the feed
pump continues to rotate (coast down) for some time, feeding water to the vessel, and circulation continues in the undamaged loop. Eventually the feed
pump flow stops and the suction of the jet pumps (which circulate liquid in the

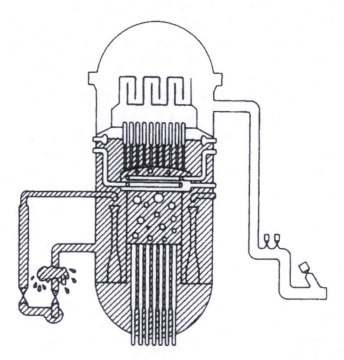

Figure 4.28: Hypothetical BWR LOCA event: time of initiation.

vessel) becomes uncovered, causing the core flow rate to drop to zero (Figure 4.29). Due to this core flow stoppage, the core begins to dry out and increase in temperature after about 10 s from the initiation of the break. The flow at the break switches mainly to steam, the water in the annular space containing the jet pumps being completely discharged, and steam formation occurs in the lower plenum as the system pressure decreases more rapidly. Vaporization occurring in this way because of depressurization is often referred to as *flashing*, and the effect of lower plenum flashing is illustrated in Figure 4.30. The flashing effect causes a two-phase mixture to flow up through the jet and the core, resulting in enhanced core heat transfer during this period.

After about 30 s, the emergency core cooling system is triggered and the automatic depressurization system operates, reducing the vessel pressure and allowing the LPCI and LPCS to come into operation. In the boiling-water reactor, the fuel is in the form of *fuel elements* consisting of a number of fuel pins mounted in a *shroud*, i.e., a rectangular box open at the upper and lower ends. The systems inject water above the core, and this water flows into the lower

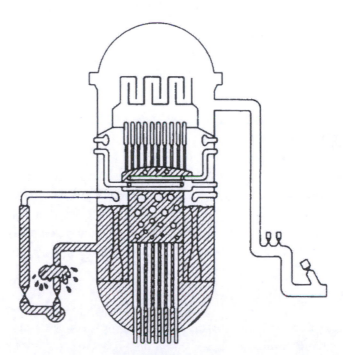

Figure 4.29: Hypothetical BWR LOCA event: time at which jet pump suction is uncovered.

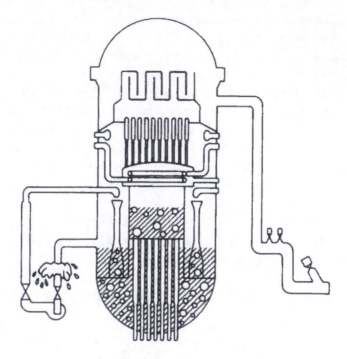

Figue 4.30: Hypothetical BWR LOCA event: lower plenum flushing.

plenum down the shroud surrounding each fuel element. The existence of this water near the fuel elements causes them to heat up much more slowly, and, eventually, water passing down the shroud into the lower plenum floods the lower plenum and water begins to rise through the core, quenching it in much the same way as in the PWR. Just as in the PWR, during this reflooding phase, the reflood rate is limited by the rate at which the generated steam can escape—the *steam binding* effect. This phase of the LOCA event is illustrated in Figure 4.31. Figure 4.32 shows a typical calculated temperature response of the shroud (channel) and fuel rods during a BWR LOCA.

4.4.2 Small-Break LOCAs in BWRs

The analysis of BWRs must also consider the full range of break sizes. Peak clad temperatures tend to be highest in the design base accident (i.e., full pipe rupture) described above. Peak clad temperature increases with break size in the

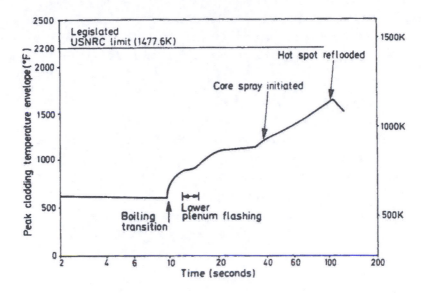

Figue 4.31: Steam binding in a BWR during a hypothetical LOCA event.

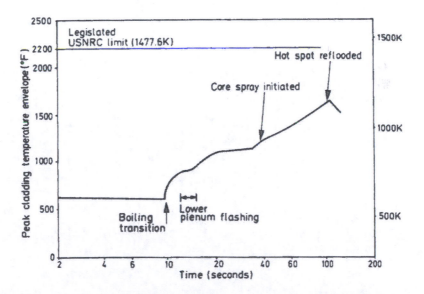

Figure 4.32: Typical BWR/6 peak cladding temperature following a design basis accident.

range 0–100 cm² and then falls with break size before rising again, reaching the value for 100 cm² again at around 1000 cm² and subsequently rising again continuously with break size up to the maximum possible size, i.e., full pipe rupture.

4.5 CANDU REACTOR

In the CANDU reactor, the coolant is distributed to and collected from the core by pipes known as *headers*, which are connected in turn to each of the fuel channels by other tubes known as *feeders*. The circuit for the CANDU reactor is illustrated schematically in Figure 4.33.

If a loss-of-coolant accident occurs, emergency coolant is injected into all the

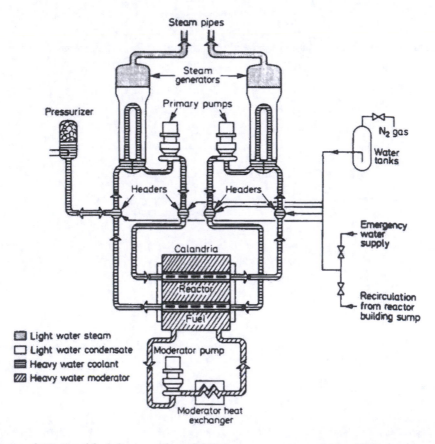

Figure 4.33: Simplified diagram of a CANDU heat transport system (and ECI system).

headers by a separate emergency coolant injection (ECI) system. This system supplies light water to the reactor during the LOCA as shown schematically in Figure 4.33. The system has a high-pressure injection stage in which gas pressure is employed to inject the water into the headers in a manner similar to the accumulators in the PWR. In some designs, this gas-pressurized system is replaced with high-pressure pumps that draw water from an emergency water tank. When the high-pressure supply is exhausted, water is pumped at a lower pressure from a separate water tank and fed into the reactor. Finally, the water being emitted from the reactor circuit into the containment building is recovered and pumped back to the headers via a heat exchanger that cools the entering water stream.

With the CANDU reactor, there are two main disadvantages related to behavior during a LOCA:

1. The reactor channels are horizontal. This means that if steam voids are formed on the channel, the water phase separates toward the bottom of the channel, leaving the top part of the channel in steam and relatively uncooled. This gravitational stratification effect is of great importance in considering the behavior of the fuel following a LOCA.

2. The CANDU reactor has a positive void coefficient; i.e., when voids are formed in the heavy-water coolant, the reactivity increases because the creation of the voids in the fuel channel makes little difference to the overall volume of moderator in the system. Thus, the neutron absorption in the heavy water in the fuel channels is removed and the reactivity increases. In a typical transient the fuel power can increase by a factor of 2 within 1 s after the accident, followed by a rapid decrease as the shutdown systems begin to operate. In view of these positive reactivity effects, it is important, for safety, to have two independent shutdown systems, as illustrated in Figure 4.34. In the first system, cadmium shutoff rods fall under gravity from the top of the reactor. In the second a neutron-absorbing solution (poison) is injected through horizontal nozzles into the heavy-water moderator surrounding the fuel channels.

Another potential problem with the CANDU reactor under LOCA conditions (with a break, say, in the inlet header) is that of *flow stagnation*. Water is sucked out of one end of the channel by the pump and leaves from the other end of the channel toward the break. The center part of the channel can therefore be stagnant, and this leads to rapid overheating of the fuel. In the design of the CANDU reactor, careful attention must be given to these potential problems. However, there are two mitigating features of CANDU reactors that help in this regard:

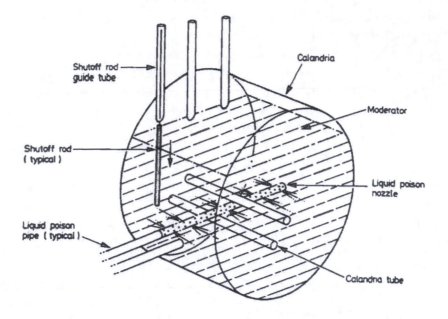

Figure 4.34: Shutdown systems: shutoff rods and liquid "poison" injection.

1. In accidents that cause the fuel and pressure tube to heat up, substantial amounts of heat can be transferred to the moderator, which can serve as an in-core heat sink.
2. Since the control rods that penetrate the cold, low-pressure moderator are operating under low-temperature conditions, it can be argued that the systems are much more reliable than those which operate at high temperature and pressure. A more detailed review of the safety of CANDU reactors is given by Snell, V.G., et al. (1990).

4.6 GAS-COOLED REACTORS

The safety of both Magnox reactors (Figure 2.4) and advanced gas-cooled reactors (AGRs) (Figure 2.5) has common elements, and the two reactor types will be dealt with together here. However, many of the detailed points are more relevant to the more modern form of gas-cooled reactor, namely, the AGR.

Using the classification of operational states outlined in Section 4.1, we regard the following forms of transient behavior as relevant:

1. *Operational transients.* Operational transients of the type encountered in water reactors—e.g., the problems of start-up and shutdown and of variation of load during operation—are also found in gas-cooled reactors. Attainment of criticality in the reactor is controlled by the operator, who is prevented by various interlocks from carrying out actions that are potentially hazardous. For instance, the control rods may not be raised until the main reactor protection system is operative. Another form of operational transients that occurs in gas-cooled reactors is associated with replacing used fuel elements with new ones (refueling) while the reactor is operating at power. Briefly, this process demands attaching a small pressure vessel to the cooling channel, breaking into the primary system and extracting the fuel element into the subsidiary pressure vessel, releasing a new fuel element from the subsidiary pressure vessel (sometimes called the *refueling machine* or *recharging machine*), and sealing the primary circuit before removing the spent fuel element for further processing.

2. *Upsets.* Again, similar upset conditions are encountered in gas-cooled reactors and water-cooled reactors. An upset can consist of loss of site power, a turbine trip, or faults on the secondary/steam side. An example specific to the gas-cooled reactor would be failure of one of the gas circulators.

3. *Emergency conditions.* Interruption of the normal electricity supply to the power station represents the emergency condition in gas-cooled reactors. An automatic reactor trip shuts down the fission reaction and is initiated by a drop in circulator supply voltage or in circulator speed. Upon loss of electrical supplies from the grid, diesel generators are brought into operation automatically to provide essential power supplies to the plant, including the circulators. Heat is extracted from the circulating gas by means of special heat exchangers known as *decay heat boilers*. The AGRs are designed such that even if it is not possible to maintain circulator rotation, natural circulation of the gas through the core and then through the decay heat boilers will be sufficient to remove the decay heat. The effectiveness of this process is illustrated in Figure 4.35. It is estimated that natural circulation flow represents about 2% of the normal full-power flow, whereas, as shown in Figure 4.35, any flow above about 0.35% of the normal flow would be sufficient to maintain the fuel temperature below the maximum allowable value of 1350°C to prevent excessive clad corrosion. Other faults leading to emergency conditions include:

   a. Boiler feedwater faults. Loss of boiler feedwater would lead to an increase in coolant gas outlet temperature from the boiler that could, if sufficiently severe, potentially damage the gas circulator. The reactor is tripped and posttrip cooling is provided by the decay heat boiler system.

   b. Steam line breaks. The AGR is divided into four quadrants, each of which has two circulators and two boilers with associated control and protection systems. Failure of a steam main from one of the boilers could,

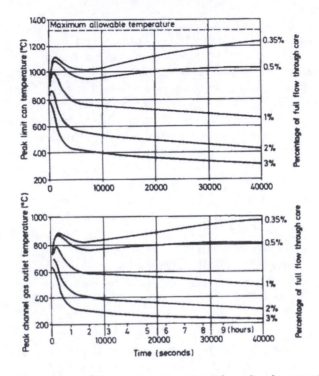

Figure 4.35: AGR temperature following a reactor trip with cooling by natural circulation.

at worst, render two quadrants of the plant unavailable. Again, the reactor is tripped and posttrip cooling is provided by the main boilers and the decay heat removal boilers.

c. Water entering the reactor. A fault in the boiler could lead to water entering the primary coolant circuit. The presence of steam arising from the boilers would give a rapid increase in pressure, causing a reactor trip. The reactor pressure vessel is protected against overpressurization by safety relief valves.

4. Limiting fault conditions. For gas-cooled reactors, typical faults in this category might be:

a. Depressurization following a breach of the primary circuit outside the prestressed concrete pressure vessel, e.g., through a stuck-open safety valve or a break in the pipework in the gas purification plant.

b. Withdrawal of a group of control rods either at power or with the reactor shut down.

c. Single-channel faults resulting from blockages or fracture of the graphite sleeves surrounding the fuel element. Of these limiting fault conditions, the depressurization fault is considered the most severe and is discussed further below.

4.6.1   Design Basis Accident for the AGR: Depressurization Fault

The heat transport capacity of carbon dioxide falls essentially in proportion to its density; in a depressurization from 40 bars to 1 bar (atmospheric pressure) the density is reduced by a factor of 40, reducing the heat transport capacity similarly. Provided the reactor is tripped as a result of the depressurization, the reduction in heat removal capacity is quite closely matched by the reduction in heat generated in the fuel in going from normal operation to shutdown (where there is only decay heat to consider). Thus, it should not be necessary for fuel temperatures to rise significantly above their normal operating values during a depressurization accident in a gas-cooled reactor.

Guaranteeing heat removal capacity after a depressurization presupposes that a means is always provided to circulate the coolant adequately. As we saw above, if the reactor is not depressurized during an emergency condition in which the circulators become inoperative, natural circulation cooling is sufficient to take away the heat. However, if the circulators are inoperative and the reactor is depressurized, natural circulation may be insufficient to keep the fuel temperatures below melting.

There are several mitigating circumstances related to depressurization and fuel temperature increase in an AGR. First, it is an integral type of circuit (see Section 3.7), and the majority of components are inside the containment vessel. Thus, the diameter of the maximum break is limited to about 200 mm. This means that the depressurization from such a large vessel (which is equivalent in volume to about 30 PWR vessels) is very slow. Typically, it might take about an hour to depressurize the vessel from its operating condition to atmospheric pressure. During this time, the decay heat rate diminishes substantially (see Table 2.2). However, even at this reduced rate, it is important to keep at least one of the circulators operational in order to maintain long-term cooling. Thus, an essential feature of safety protection in an AGR is that of safeguarding the integrity of operation of the circulators. This is achieved by having diversified backup electricity supplies to ensure that power is available to drive the circulators together with reliable supplies of water to cool the oil, which is used both as the circulator seals and for circulator cooling.

Another safety problem related to an AGR is that the prestressed concrete pressure vessel must be maintained at all times at a temperature less than 100°C. This condition is achieved in normal operation by using cooling water pipes set into the concrete vessel walls. In handling fault conditions it is important to maintain this cooling water supply, and this is done by having auxiliary and reliable supplies available on site. Auxiliary supplies are also needed to ensure that the feedwater to the decay heat boilers is always maintained.

Finally, the need for assured supplies of electricity, cooling water, and feed-water means that very great care must be taken to provide a diversity of sup-plies in case of failure. For instance, there must be at least four sets of diesel generators to provide electricity for the circulators. The safety of the reactor would be assured if only one of these was available.

4.7   SODIUM-COOLED FAST REACTOR

The various operational states for a liquid metal–cooled fast reactor (LMFBR) can be listed as follows:

1. *Normal operation and operational transients*. The sodium in the circuit is al-ways kept in a molten state by heating the whole circuit with electrical resis-tance heaters wound on all the pipework. This maintains the sodium at a temperature of at least 100°C (the melting point of sodium is 98°C). The large pool of molten sodium responds rather slowly to heat input. Thus, the coolant takes some time to reach operating temperature.

2. *Upsets*. Various categories of upset situations have been postulated for an LMFBR. Many are similar to those for water- and gas-cooled reactors, includ-ing loss of load, turbine trip, loss of feedwater, and loss of a single main cir-culating pump.

3. *Emergency conditions*. In an LMFBR, emergency conditions will occur if the upsets described above cannot be contained within normal operational pro-cedure. These include the following:

   a.   Loss of electric power (and resultant coast-down of the pumps). Loss of power supply to the primary coolant pumps causes them to coast down to zero speed. Under these circumstances, the reactor is immediately tripped and power may be reinstated to the circulators from emergency supplies (diesel-driven generators) that operate secondary electric motors ("pony" motors). However, the sodium pool itself represents a major heat sink. For instance, with the decay heat in the reactor alone, the sodium pool would take about 24 h to reach the boiling point if there were no heat removal at all. Moreover, the reactor has decay heat removal heat exchangers that are connected to the primary circuit and can remove the decay heat by natural circulation alone, without any electric power input to the reactor. The final heat sink from these removal systems is the atmosphere via air-cooled heat exchangers. Even if single-phase natural circulation is not established immediately after a reactor trip, sodium boiling in the core is an acceptable means of removing decay heat and the generation of two-phase flow within the core enhances the natural circulation to the extent that single-phase natural circulation is rapidly established.

b.   Inadvertent increase in neutron population in the core. The rate of the fission reaction in the LMFBR can be increased by inadvertent removal of a control rod, movements of the fuel (e.g., by the fuel elements becoming bowed, as happened in the U.S. experimental breeder reactor EBR I incident described in Chapter 5), or sodium boiling in the core. Sodium boiling in the inner region of the core causes an increase in the rate of fission (neutron population), since sodium absorbs neutrons, and if it is partially vaporized, the absorption is reduced. However, if the boiling occurs in the outer region of the core, the reduced local density causes increased leakage of neutrons from the core and gives rise to a reduction in the fission reaction (neutron population). Thus, the effect of sodium boiling is usually negative for small reactors such as the prototype fast reactor (PFR) and positive for larger reactors, where any boiling is likely to be away from the boundary of the core. Great care must be taken to design LMFBRs to avoid failure in the control rod insertion mechanism, and systems are being designed to be capable of self-actuated shutdown, directly triggered by high temperatures in the core and requiring no out-of-reactor mechanisms.

c.   Local damage within a fuel subassembly. The reactor core consists of hundreds of separate groups of fuel elements, which can be inserted or removed independently from the core. A typical subassembly consists of 300 pins 6 mm in diameter and I m long. Since an accident in the Enrico Fermi reactor (described in Chapter 5), considerable attention has been focused on the possibility of blockages occurring within individual subassemblies or groups of subassemblies. If the sodium flow is blocked, local melting of the cladding and possibly the oxide fuel could occur. The oxide fuel reacts with the sodium, limiting its useful lifetime, but the failure of a subassembly can usually be detected by specially provided instrumentation. Failure to detect the fault may lead to escalation of the upset into a fault condition (see below), with debris blocking an increasing area of the core, reducing the flow, and preventing cooling. Reduction of flow gives local sodium boiling, and this increases the reactivity in the region, making the problem worse.

d.   Loss of heat removal from secondary sodium or steam systems. Here the system responds in the manner described for the loss-of-flow upset. The reactor is tripped and natural circulation cooling is set up, with heat released by the decay heat removal heat exchangers. The circulators may still operate under these circumstances; provision is made for driving them automatically via the pony motors.

To summarize, the primary objective in the design and operation of an LMFBR is to bring it, in response to the various operating states, to a condition such that it can be cooled by either (1) primary circuit cooling by the intermediate sodium circuit to the steam generators or (2) primary circuit cooling via the separate liquid metal coolant circuit to air-cooled heat exchangers. In the former case, the primary circuit uses forced circulation, while the secondary intermediate circuit can rely on natural convection. Emergency boiler-feed is supplied to the steam generators, and the steam produced vented from the circuit. In the latter case, natural circulation in the primary circuit is sufficient to cool the core, and indeed if all the heat exchangers are operational, natural convection is sufficient in the secondary circuit. However, if this is not the case, a powered fan is necessary to force air across the heat exchanger.

For the fast reactor much attention has been given to the case that is beyond the design basis accident, namely, conditions under which quantities of molten fuel are produced. In this case it is postulated that the energy present in the molten fuel could be rapidly converted to a pressure shock wave and cause a vapor explosion. We shall consider this extremely unlikely event in Chapter 6.

REFERENCE

Snell, V.G., et al. (1990): "CANDU Safety under Severe Accidents: An Overview." *Nuclear Safety* 31 (January–March): 20–35.

EXAMPLES AND PROBLEMS

1 LOCA in a PWR

Example: A major loss-of-coolant accident occurs in a PWR. The reactor is tripped and goes subcritical after 1 s; dryout (see Section 3.3) occurs after 4 s when the heat transfer coefficient from the fuel pin drops from 50,000 W/m^2K to a very low value; the blowdown is complete in 30 s. The fuel pins consist of 10-mm UO_2 fuel pellets in an 11-mm outside diameter Zircaloy can. The maximum rating R is 40 kW/m. The coolant is initially at 300°C; the temperature drops through the can and across the fuel-pellet-to-can gap are initially 50 K and 300 K, respectively. What is the can temperature at the end of the blowdown phase?

After 30 s the ECCS system operates and provides water at 30°C. The heat transfer coefficient during the refill stage is 50 W/m^2 K. What is the maximum temperature recorded during the refill stage, and when is it recorded?

Solution: There are four sources of energy that could make the can temperature rise.

(1) *Delay in shutting down the reactor.* We will assume that from the start of the transient to the time when the reactor goes subcritical at 1 s, normal cooling is provided and this component makes no contribution to can temperature rise.

(2) *Internal energy stored in the fuel.*

(a) Fuel pellet temperatures equalize at a time t, given by $s\,a^2/\kappa$, where s is a shape factor (equal to 0.2 in this case), κ is the thermal diffusivity ($= k/\rho c_p$), and a the fuel pellet radius. For UO_2, $k = 2.5$ W/m K, $\varrho = 10,000$ kg/m^3, and $c_p = 350$ J/kg K. Therefore, $\kappa = 0.7 \times 10^{-6}$ m^2/s, and for a 5-mm fuel pellet temperatures equalize in

$$t\left(=\frac{0.2\times25\times10^{-6}}{0.7\times10^{-6}}\right)=7.1\,\text{s}$$

(b) The thermal capacity of the fuel per unit length (C_f) is given by

$$C_f = \pi a^2 \varrho c_p$$
$$= \pi \times 25 \times 10^{-6} \times 10^{-4} \times 350$$
$$= 275 \, \text{J/m K}$$

The thermal capacity of the can (C_c) is given by

$$C_c = \pi(b^2 - a^2)\varrho c_c$$

where b is the outer can radius, ϱ_c is the can density (6500 kg/m^3), and c_c is the can material specific heat (350 J/kg K). Thus

$$C_c = \pi \times (0.0011^2 - 0.0010^2) \times 6500 \times 350$$
$$= 350 \, \text{J/kg K}$$

(c) The energy stored in the fuel pin *above the can surface temperature* is made up of two parts: that in the fuel pellet and that in the can.

(3) *Stored energy is the fuel pellet.* This is the product of the thermal capacity of the fuel per unit meter and the mean temperature of the fuel pellet above the can surface temperature (T_s) Since the time (30 s) is long compared with the time for equalization across the fuel pellet, the fully equalized pellet temperature can be used and is

$$\frac{R}{8\pi k}\left(=\frac{40,000}{8\times\pi\times2.5}\right)=637\,\text{K}$$

To this must be added the temperature drop across the can and across the pellet/can gap, 50 K + 300 K = 350 K.

Therefore, stored energy in the pellet *above* $T_s = C_f(637 + 350) = 2.71\times10^5$ J.

(4) *Stored energy in the can.* This is the product of the thermal capacity of the can per unit meter and the mean temperature of the can above the can surface temperature T_s
We take the mean temperature as half the temperature rise across the can ($= 50/2$) = 25 K. Therefore, the stored energy in the can *above* $T_s = C_c \times 25 = 937$ J (which can be neglected in relation to the stored energy of the fuel pellets).

(5) *Residual fission heating*—as the neutron chain reaction dies away. This can be taken as

$$1.6 \times R = 1.6 \times 40,000 = 64,000 \text{ J}$$

(6) *Fission product decay heating.* Integrating the fission product decay heating for 29 s, we have

$$1.5 \times R = 1.5 \times 40,000 = 60,000 \text{ J}$$

(7) But some heat is removed by good cooling between I and 4 s. This is estimated at 120,000 J. If we now add the various energy contributions, we have

$$(1) + (3) + (5) + (6) - (7)$$
$$0 + 271 \text{ kJ} + 64 \text{ kJ} + 60 \text{ kJ} - 120 \text{ kJ} = 275 \text{ kJ/m}$$

The thermal capacity of the fuel (pellet and can) per meter length is

$$275 + 37.5 = 312.5 \text{ J/m K}$$

Therefore, the temperature rise is 880 K.

The initial can temperature is

$$300°C + \frac{40,000}{\pi \times 0.011 \times 50,000} = 323.1°C$$

So at the end of the blowdown phase the final can temperature is 880 K + 323.1°C = 1203.1°C. However, the coolant temperature does not remain at 300°C, and during the blowdown the saturation temperature falls to that for I atm (~100°C). So the actual can temperature will be significantly less than 1203.1°C (by as much as 200°C, i.e., around 1000°C) (see also Figure 4.1 for an approximate solution to this problem).

At the end of the blowdown phase, the decay heating is 0.04 R or 1600 W/m. The heat removed by the ECCS is

$$\pi \times 0.011 \times 50 \times (1000 - 30) = 1676 \text{ W/m}$$

So the heat removal is about the same as or slightly higher than the heat released, and the can temperature will be at a maximum of 1000°C at about 30 s and will start to fall as the decay heat falls.

Problem: Repeat the calculations described in the example, but assume that dryout of the fuel occurs after 1 s (simultaneous with the reactor trip).

2 Inlet pipe rupture in a Magnox reactor

Example: A severe accident in a Magnox reactor contained in a steel pressure vessel is rupture of an inlet cooling duct followed by 50 s of stagnation in the core. During this period the only means of cooling is heat lost by radiation to the graphite moderator, which remains at 350°C. The metal fuel pin has a diameter of 30 mm, and the initial can temperature is 450°C. The temperature drops across the Magnox cladding and fuel-to-clad gap may be neglected. The initial fuel rating (R) is 35 kW/m, and it takes 4 s for the control rods to enter the reactor to shut it down. What is the maximum Magnox cladding temperature at the end of the stagnation period?

Solution: We consider the four sources of energy that will make the clad temperature rise.

(1)   *Delay in shutting down the reactor.* The energy per meter length due to the

delay in shutting down the reactor is

$$35,000 \text{ J/ms} \times 4 = 140,000 \text{ J/m}$$

(2) *Internal energy stored in the fuel.*

(a) The metal fuel temperatures equalize at a time t given by sa^2/R where R is the thermal diffusivity $(=k/\varrho c_p)$, s a shape factor $(= 0.2)$, and a the free radius. For uranium metal $k = 32$ W/m K, $\varrho = 19,000$ kg/m^3, and $c_p = 170$ J/kg K. Therefore, $k = 9.8 \times 10^{-6}$m^2/s, and for $a = 15$ mm, metal fuel temperatures equalize in

$$t = \frac{0.2 \times 225 \times 10^{-6}}{9.8 \times 10^{-6}} = 4.5 \text{ s}$$

(b) The thermal capacity of the fuel per meter length C_f is given by $\pi a^2 \varrho c_p = \pi$ x 225 x 10^{-6} x 1.9 x 10^4 x 170 = 2283 J/mK.

For the clad we are not told the mass or the dimensions of the can. From reference sources we established that the mass of clad per unit length is 1 kg/m and that c_c is 1200 J/kg K. Therefore, the thermal capacity of the can=

$$1 \times 1200 = 1200 \text{ J/m K}$$

(c) The energy stored in the fuel pin above the clad surface temperature is made of two parts: that in the metal fuel and that in the cladding (which we neglect). *The stored energy in the metal fuel* is the product of the thermal capacity of the fuel per unit meter and the mean temperature of the metal fuel above the clad surface temperature. Since the time $(50 - 5)$ s is long compared with the time for equalization across the fuel, the fully equalized metal temperature can be used and is

$$\frac{R}{8\pi K} = \left(\frac{35,000}{8 \times \pi \times 32} \right) = 43.5 \text{K}$$

We can neglect any other temperature drop in the fuel element; therefore, the stored energy in the fuel above the external clad temperature is

$$43.5 \times 2283 = 99.310 \text{ J/m}$$

(3) *Residual fission heating*—as the neutron chain reaction dies away. This can be taken as

$$1.6 \times R = 1.6 \times 35,000 = 56,000 \text{ J/m}$$

(4) *Fission product decay heating.* Integrating the fission product decay heating for $(50 - 5)$ s, we have

$$2.1 \times R = 2.1 \times 35,000 = 73,500 \text{ J/m}$$

If we now add the various energy contributions, we have

$$(1) + (2) + (3) + (4) =$$
$$140,000 + 93,310 + 56,000 + 73,500 = 362,810 \text{ J/m}$$

The thermal capacity of the fuel (metal and cladding) per unit meter is

$$2283 + 1200 = 3843 \text{ J/m K}$$

Therefore, the temperature rise is 104.1 K.

This is the temperature rise assuming no cooling of the fuel. Let us now calculate the effect of radiation of heat to the moderator. The moderator is assumed to be at a temperature of 350°C; the fuel element is assumed to be a cylinder of 50 mm (fin outside diameter) in a fuel channel of 100 mm.

The average clad temperature during the transient is

$$450 + \frac{104}{2} = 502°C$$

The heat lost by radiation may be calculated from the formula

$$\frac{\alpha A_1 (T_1^4 - T_2^4)}{(1/\varepsilon_1) + (A_1 + A_2)(1/\varepsilon_2 - 1)}$$

where ε_1 and ε_2 are the emissivities of the fuel and graphite, respectively, and are assumed to be 0.6 in each case. $\sigma = 5.67 \times 10^{-8}$ W/m^2 K^4 (Stefan's constant), and the surface areas of the fuel element (A_1) and the graphite channel (A_2) per meter are

$$A_1 = 0.157 \text{ m}^2$$
$$A_2 = 0.314 \text{ m}^2$$

(This assumes a 5-cm-diameter can in a 10-cm-diameter channel.) Heat lost by radiation is thus

$$\frac{5.67 \times 10^{-8} \times 0.157 (775^4 - 623^4)}{1.98} = 945 \text{ W/m}$$

The decay heat rate at (50 – 5) s is 0.038 of full power = 0.038 x 35,000 = 1330 W/m. So the heat lost by radiation is *less* than the decay heat rate, and in the absence of any other cooling the fuel element will heat up at

$$\frac{1330 - 945}{3483} = 0.11 \text{ K/s}$$

until the heat removed by radiation matches the decay heat rate.

The total heat removed during 50 s by radiation is

$$945 \times 50 = 47,250 \text{ J/m}$$

corresponding to a reduction of 13.4 K compared with the no-cooling temperature rise of 104.1 K.

Therefore, the maximum cladding temperature is

$$450 + 104.1 - 13.5 = 540.6°C$$

Problem: A new Magnox fuel design is being considered in which the fuel element diameter is to be increased from 5 cm to 6.5 cm and the uranium metal fuel pin diameter is increased to 4 cm. Repeat the calculations given in the example, and evaluate the effect of this design change on the temperatures reached in the specified accident.

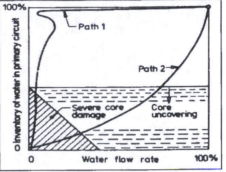

3 Pumps on or pumps off?

Problem: There has been considerable controversy about whether the operators should leave the main circulating pumps operating or stop them during a small break loss-of-coolant accident for a PWR. With the aid of the accompanying diagram, discuss the events that occur when the pumps are stopped (path 1) or left operating (path 2).

BIBLIOGRAPHY

Bankoff, S.F., and N.H. Afgan (1982). *Heat Transfer in Nuclear Reactor Safety.* Proceedings of a Seminar on Nuclear Reactor Safety Heat Transfer, Dubrovnik, Yugoslavia, September 1–5, 1980. Hemisphere, Washington, D.C., 964 pp.

Barsell, A.W. (1981). *A Study of the Risk Due to Accidents in Nuclear Power Plants.* German Risk Study—Main Report, various pages.

Bergles, A.E., J.G. Collier., J.M. Delhaye, G.F. Hewitt, and F. Mayinger, (1981). *Two-Phase Flow and Heat Transfer in the Power and Process Industries.* Hemisphere, Washington, D.C., 719 pp.

Bowerman, J.M., and G.C. Dale, eds. (1982). *The Safety of the AGR.* CEGB and SSEB, 111 pp.

Fast Reactor Safety Technology. Proceedings of a Meeting, Seattle, August 19–23, 1979. American Nuclear Society, LaGrange Park, Ill., 5 vols.

Flowers, B. (1976). *Nuclear Power and the Environment.* Royal Commission on Environmental Pollution, Sixth Report. HMSO, London, 237 pp.

Jones, 0.C. (1981). *Nuclear Reactor Safety Heat Transfer.* Papers presented at the summer school on nuclear reactor safety heat transfer, Dubrovnik, Yugoslavia, August 24–29, 1980. Hemisphere, Washington, D.C., 959 pp.

Judd, A.M. (1981). *Fast Breeder Reactors: An Engineering Introduction.* Pergamon, Elmsford, N.Y., 161 pp.

Menlo. M. (1983). *Thermal-Hydraulics of Nuclear Reactors.* Papers presented at the Second International Topical Meeting on Nuclear Reactor Thermal-Hydraulics, Santa Barbara, Calif., January 11–14, 1983, 1529 pp.

5

Loss-of-Cooling Accidents:
Some Examples

5.1 INTRODUCTION

Incidents at nuclear power stations create a great deal of public interest and sometimes concern and alarm.

Many incidents occurred in the 50-year period up to 1995, though very few resulted in injury or death to plant operators or the general public. Mosey (1990) lists 60 or so separate events, and even this list is probably not comprehensive. The three most serious events are the accidents at Windscale (1957), Three Mile Island (1979), and Chernobyl (1986). Brief details of each are included in this chapter.

Designers of nuclear power stations do assess the risks and consequences beyond the basis adopted for design. However, prior to Chernobyl, the actual release of radioactive fission products from nuclear accidents had been very much less than predicted from such analyses, indicating their general conservatism. Accidents can be examined by looking at which of the three basic safety principles, the Three Cs—*control the reaction, cool the fuel,* and *contain the radioactivity,* has been breached and to what extent the overall *defense in depth* has been challenged.

If the world is to benefit from nuclear energy in the longer term despite the potential dangers involved, it is essential that the lessons learned from each accident or incident are incorporated into future designs and into operator training and safety management to make existing stations safer. It is beyond the scope of this book to examine all these incidents. Rather, we will select those examples which illustrate specific points we have highlighted in previous chapters. The examples are chosen to illustrate both type of fault and type of reactor, as follows:

Light water–cooled reactors	*Gas-cooled reactors*
SL-1	Windscale
Millstone 1	St. Laurent
Browns Ferry 1 and 2	Hunterston B
Three Mile Island-2	Hinkley Point B
Ginna	*Liquid metal–*
Mihama-2	*cooled reactors*
Chernobyl	EBR-1
Heavy water–cooled reactors	Enrico Fermi
NRX	
Lucens	

5.2 INCIDENTS IN LIGHT WATER–COOLED REACTORS

5.2.1 The SL-I Accident

A small (thermal capacity, 3 MW) experimental boiling-water reactor called SL-1 (Stationary Low-Power Plant No. 1), installed at the U.S. National Reactor Testing Station (NRTS) in Idaho, was destroyed on January 3, 1961, as a result of the manual withdrawal of a control rod while the reactor was shut down. The reactor had been shut down for maintenance and to install additional instrumentation. This work was completed during the day shift on January 3, and it was the job of the three-man crew of the 4–12 P.M. shift to reconnect the control rods. The installation of the additional instrumentation required disconnecting the control rods, leaving them fully immersed in the reactor. However, when they were disconnected, the rods could be lifted out manually. Lifting the control rods by about 40 cm (16 in.) was sufficient to make the reactor critical.

At 9:01 P.M. on January 3, alarms sounded at the fire stations and security headquarters of the NRTS, which was located some distance from the SL-1 facility. Upon investigation it was found that two operators had been killed (a third died later) and that high radiation levels were present in the building. The exact reason for the accident has never been discovered; the removal of the control rod could have been accidental or deliberate, but no one will ever know.

Based on a careful examination of the remains of the core and the vessel during the cleanup phase, it was concluded that the control rod had been withdrawn by about 50 cm (20 in.), sufficient for a very large increase in reactivity. The resulting power surge caused the reactor power to reach 20,000 MW in about 0.01 s. This caused the plate-type fuel to melt. The molten fuel interacted with the water in the vessel, and the explosive formation of steam caused the water above the core to rise with such force that when it hit the lid of the pressure vessel, the vessel itself rose 3 m (9 ft) in the air and then dropped back approximately to its original position.

Two main lessons were learned from this incident:

1. It is unsatisfactory to have any reactor system (even a small experimental reactor of this kind) in which removal of control rods is not prevented by a suitable series of interlocks. Removal of a control rod as in the SL-1 accident would be impossible in a modern power reactor.

2. Ejection of water from the core normally leads to a decrease in reactivity, which automatically shuts down the reactor by additional void formation. However, as the SL-1 accident showed, a very fast increase in reactivity can melt the fuel before significant voids are formed to shut down the fission reaction. This effect was demonstrated deliberately in another U.S. reactor test: the so-called BORAX reactor was deliberately brought into this condition and destroyed in 1954.

Explosions arising from the interaction of molten fuel and liquid coolant will be discussed further in Chapter 6.

5.2.2   The Millstone 1 Accident

On September 1, 1972, a routine start-up operation was proceeding on the Millstone 1 reactor in Connecticut. This reactor was a 660-MW(e) BWR. When the reactor had achieved less than 0.1% of full power, the operator noted that the water purification system was malfunctioning. He switched to a second water purification system and continued the start-up. About half an hour later the second system also failed and the operator began to shut down the reactor. When it became obvious that salt from seawater was penetrating the primary coolant circuit, the reactor was tripped rapidly. Upon investigation it was found that tubes in the condenser (which was cooled by seawater) had corroded, allowing a massive amount of seawater to enter the primary circuit. One consequence of the saltwater ingress was the failure of the instruments that measured power in the reactor; the failure was due to stress corrosion cracking of the stainless steel sheaths of the instruments, which are sensitive to chloride attack.

The reactor was successfully repaired and resumed operation. Although this accident caused no injuries and no radioactivity was released, it demonstrates the relative vulnerability of direct-cycle systems such as the BWR in comparison with indirect-cycle systems such as PWR, CANDU, or AGR. In a BWR the primary circuit coolant passes directly to the turbine and is condensed in the condensers before returning to the reactor. If the condensers are cooled by seawater, ingress into the primary circuit is always a potential problem. One way to overcome this is to isolate the condensers in the event of leakage of seawater, but this leads to loss of the main heat sink and a need to provide alternative cooling or means of energy release.

5.2.3 The Browns Ferry Fire

The Browns Ferry nuclear power plant in Alabama consists of three 1065-MW(e) boiling-water reactors. On March 22, 1975, a workman who was lying on his side used a lighted candle to test for leakage of air around cable penetrations through a concrete wall at the Unit I plant. A hole was found, and the workmen stuffed some polyurethane sheet into it and tested again for leaks. The leak persisted, and the candle flame ignited the polyurethane sheet. The air rushing into the hole spread the fire into the hole and away from the workmen, so that they could not extinguish it with fire extinguishers. The fire burned for 7 h before it was put out. Units I and 2 were both at full power when the fire started. (Unit 3 was under construction and was not affected by the accident.) The fire spread horizontally and vertically, affected about 2000 cables, and caused damage that cost about $10 million to repair. There was a reluctance to use water on the fire until both reactors were in a stable shutdown condition because of the possibility of short-circuiting. Once water was used, the fire was rapidly put out.

Both reactors were shut down. However, because of the fire, both the shutdown cooling system and the emergency core cooling system for Unit 1 were inoperable for several hours. The operators had to use alternative means of injecting water into the reactor, which included a pump used in connection with the control rod drive system and pumps used for returning condensate to the system. The use of these alternative water supplies required depressurization of the reactor, and during this maneuver, the water level over the core dropped to 1.2 m above the top of the fuel. However, sufficient cooling was provided throughout the incident to prevent the core from overheating. No significant problems were encountered with the cooling of Unit 2, and the high-pressure cooling system

(HPIS) was successfully initiated. There was no release of radioactivity off-site, and no one on the site was seriously injured. Both units were, however, out of operation for over 1 year while the damage was repaired.

The main lesson from the Browns Ferry incident was related to what is called *common mode failure*. All the cables related to the safety systems passed through a single duct and failed in a common mode (despite the diversity introduced as discussed in Chapter 4), and all the systems failed when there was a fire. The moral is that the designer should ensure that each of the independent systems is truly independent and that supplies and controls to the instrumentation and actuation devices should not pass along common ductwork. The technical term for this is *segregation*, and after the Browns Ferry incident the provisions for segregation were significantly improved. For example, 3-h fire-resistant physical barriers are now placed between components, and when this is not possible the cables are separated by significant distances (typically 7 m) and protected by active fire-fighting equipment so that the possibility of a fire spreading from one to another is remote.

5.2.4   The Three Mile Island (TMI) Accident

The worst accident in the United States happened in March 1979 at the No. 2 reactor at the Three Mile Island nuclear plant near Harrisburg, Pennsylvania. The plant consists of two Babcock & Wilcox pressurized-water reactors, each having an electrical capacity of 961 MW(e).

At about 4 A.M. on March 28, 1979, a condensate pump moving water from the condensers in the turbine building stopped. This led to tripping of the main steam generator feedwater pumps (which would otherwise have been starved of water), which in turn led to the turbine's being tripped. As we saw in Chapter 4, this is a normal upset condition, and the incident should have proceeded benignly according to the design. To see why this did not happen, it is helpful to examine each phase of the accident in turn.

Phase 1. Turbine Trip (0–6 mm).   This phase is illustrated in Figure 5.1. The valves that allow steam to be dumped to the condenser opened as designed and the auxiliary feedwater pumps started. The interruption of the flow of feedwater to the steam generators caused a reduction in heat removal from the primary system. The reactor coolant system responded to the turbine trip in the expected manner. The reactor coolant pumps continued to operate and to

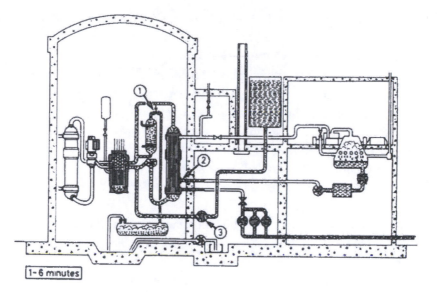

1- 6 minutes

Figure 5.1: TMI-2 phase1: turbine trip.

maintain coolant flow through the core. The reactor coolant system pressure
started to rise because the heat generated by the core—which was still operating—was not being removed from the system at the required rate by the steam
generators. This rise in system pressure caused the power-operated relief valve
(PORV) on top of the pressurizer (1 in Figure 5.1) to operate to relieve the pressure. However, the opening of this valve was insufficient to reduce the pressure
immediately, and the pressure continued to increase. The operation of the valve
occurred between 3 and 6 s after the turbine trip, and the pressurization continued until 8 s after the start of the incident, when the primary circuit pressure
reached 162 bars. At this point the control rods were automatically driven into
the core as a result of a protection system signal's detecting the overpressurization. This immediately stopped the fission reaction. At this early stage all the automatic protection features had operated as designed, and the reactor had been
shut down. However, as we explained in previous chapters (and as indicated in
Table 2.2), the decay heat remains significant. Under normal circumstances this
can be dealt with straightforwardly by the various coolant systems.

At 13 s the now-decreasing coolant pressure reached the set point for automatic closure of the PORV. The *valve failed to close*, and this first departure from
the expected response changed the incident from an upset into an emergency

event, as defined in Chapter 4. The sequence that started at this stage was very similar to the small-break accident described in Section 4.3.4. Coolant circuit water was being lost through the stuck-open PORV. In the secondary circuit, all three auxiliary feedwater pumps were running, yet the water level in the steam generators was continuing to fall and they were drying out. The reason for this was that no water was actually being injected into the steam generators because of closed valves between the auxiliary pumps and the steam generators. The valves had been closed some time before the incident (probably at least 42 h earlier) for routine testing and had *apparently been inadvertently left in that position*. The warning lights indicating the valve closure had been obscured by tags on the control board.

Thus, during this first crucial period, the reactor coolant circuit was deprived of an effective means of heat removal and could only dispose of the energy by blowing off water and steam. As we saw in Chapter 4, this was an inadequate heat removal method. One minute after the incident, the difference in temperature between the hot and cold legs of the primary circuit was rapidly reaching zero, indicating that the steam generators were drying out. The reactor circuit pressure was also dropping. At about this time the liquid level in the pressurizer began to rise rapidly. At 2 min 4s the reactor circuit pressure had dropped to 110 bars, and the emergency core cooling system (ECCS) triggered automatically, feeding cold borated water into the primary coolant system. The liquid level in the pressurizer was continuing to rise. Concern was expressed that the HPIS was increasing the water inventory in the primary circuit and that the steam above the water level in the pressurizer would be lost, preventing efficient pressure control. In effect, the system would then be full of water. Subsequent analysis has shown that, initially, expansion of the water as it heated up and, later, boiling in parts of the circuit displaced water into the pressurizer, causing the increase in pressurizer level. Because of their concern about the pressurizer level and their belief that the HPIS system was filling it, the operators tripped (shut off) one of the HPIS pumps at 4 min 38 s; the other pumps continued to be operated in a partly closed condition.

Phase 2: Loss of Coolant (6–20 min). At 6 min the pressurizer was completely filled with water. The reactor drain tank (item 7 in Figure 5.2) started to pressurize rapidly, and at 7 min 43 s the reactor building sump pump switched on to transfer water from the sump to the various wastewater tanks located in the auxiliary building. Thus, slightly radioactive water was being transferred out of the containment into the auxiliary building.

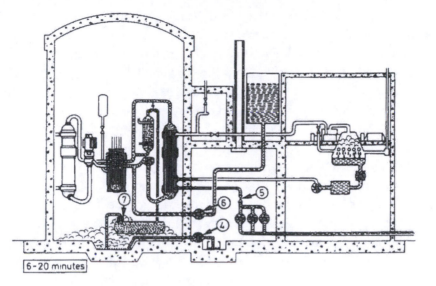

Figure 5.2: TMI-2 phase 2: loss of coolant.

In the Babcock & Wilcox TMI design, the automatic closure of valves linking the containment to the auxiliary building was not initiated unless the reactor building pressure exceeded 270 millibars. In reactors supplied by other vendors, control systems close off these connecting lines automatically when the ECCS system is actuated.

After 8 min the operators found that the steam generators were dry. Checks showed that the auxiliary feed pumps were running but that the valves were shut. The operator opened the valves, allowing feedwater to pass into the steam generators, and the reactor circuit water temperature started to drop as a result. "Hammering" and "crackling" were heard from the steam generators, confirming that the auxiliary feed pumps were now delivering water to them. The closed valves in the auxiliary feedwater circuit received a great deal of publicity immediately after the accident. It now seems likely that the unavailability of the auxiliary feedwater for the first 8 min of the accident did not, in the event, significantly affect the future course of the accident, which was largely determined by the stuck-open PORV

At 10 min 24 s, a second HPIS pump (item 6, Figure 5.2) tripped out, was restarted, but tripped out again, to be eventually restarted at 11 min 24 s, but in a throttled condition. The balance between the flow of water into the reactor from the HPIS and the flow out of the reactor from the PORV was such that there was a net outflow from the primary cooling system. At about 11 min, the pressurizer level indication was back on scale and the level was decreasing. At 15 min, the reactor coolant drain tank bursting disk (item 7, Figure 5.2) ruptured and hot water flashed into the containment building, giving a rise of pressure within that building. The coolant was now being discharged from the primary circuit, was emptying into the containment, and was passing from the containment sump, through the sump pump that continued to operate, into the auxiliary building.

At 18 min, there was a sharp increase in activity measured by the ventilation system monitors. This activity arose from the discharge of the slightly radioactive primary coolant into the containment and not from any fuel failures at this stage. At this point, the reactor circuit pressure was only about 83 bars and falling.

Up to this stage, the events at TMI-2 were very similar to a feedwater transient experienced at the Davis-Besse plant at Oak Harbor, Ohio, in September 1977. At Oak Harbor, however, the operators recognized after 21 min that the PORV had stuck open, and they closed its associated block valve, thus ending the incident. The block valve is in series with the PORV and can be manually operated to seal this line.

Phase 3: Continued Depressurization (20 min–2 h). Between 20 min and 1 h, the system parameters were stabilized at the saturation condition, about 70 bars and 290°C. At 38 min the reactor building sump pumps were turned off after approximately 30 m^3 of water had been pumped into the auxiliary building. The amount of radioactivity transferred was relatively small, since the transfer was stopped before any significant failure of fuel occurred.

At 1 h 14 min, the main reactor coolant pumps in loop B (one of two loops in the reactor—each loop has two coolant pumps) were tripped because of indications of high vibration, low system pressure, and low coolant flow. The operators would normally be expected to take such action to prevent serious damage to the pumps and associated pipework. However, turning off the pumps in loop B allowed the steam and water phases in that circuit to separate, effectively preventing further circulation in that loop.

At 1 h 40 min, the reactor coolant pumps in loop A were tripped for the same reasons (see item 8 in Figure 5.3). One concern was that a pump seal failure could occur. The operating staff expected natural circulation of the coolant, but because of the separated steam voids in both loops, this did not take place. Subsequent analysis showed that about two-thirds of the water inventory in the

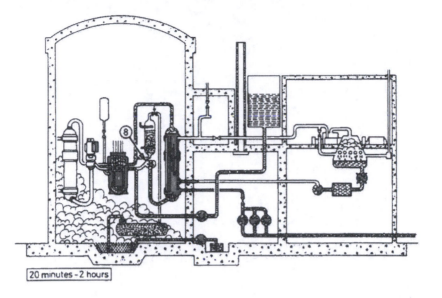

20 minutes - 2 hours

Figure 5.3: TMI-2 phase 3: continued depressurization.

primary circuit had been discharged by this stage and that when the main coolant pumps were switched off, the water level in the reactor vessel settled out about 30 cm above the top of the core. The decay heat from the core rapidly evaporated the water and brought the level down inside the core, and the core began to heat up. This overheating was the precursor of core damage.

Phase 4: The Heat-Up Transient (2–6 h).   At 2 h 18 min into the incident, the PORV block valve (item 9 in Figure 5.4) was closed by the operators. The indications of the position of the PORV were ambiguous to the operators. The control panel light indicated the actuation of a solenoid that should have closed the valve; there was no direct indication of the valve stem position. However, it must be said that failure to recognize that there had been a massive loss of reactor coolant as a result of the stuck-open PORV was the significant feature of the accident. Even at this point, however, a repressurization of the reactor coolant circuit using the HPIS would probably have successfully terminated the incident.

Following closure of the block valve, the reactor circuit pressure began to rise. At 2 h 55 min, a site emergency was declared after high radiation fields were measured in the line connecting the reactor coolant circuit to the purifica-

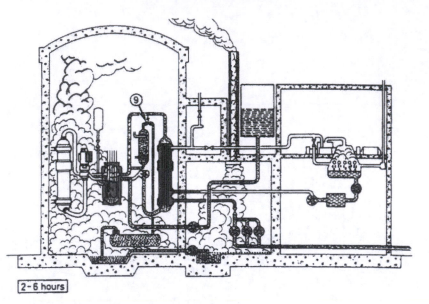

2–6 hours

Figure 5.4: TMI-2 phase 4: the heat-up transient.

tion system. By this time a substantial fraction of the reactor core was uncovered and had sustained high temperatures. This condition resulted in fuel damage, release of volatile fission products, and generation of hydrogen as a result of the interaction between the Zircaloy fuel cans and steam at high temperature.

Attempts were made to start the main reactor coolant pumps around this time. One pump in loop B did operate for 19 min but tripped out due to cavitation and vibration. The peak fuel temperature (in excess of 2000°C) was reached shortly after 3 h into the incident. At 3 h 20 min, reactivation of the HPIS effectively terminated the initial heat-up transient, both quenching the fuel and recovering the core.

A general emergency was declared about 3 h 30 min after the start of the incident as a result of rapidly increasing radiation levels in the reactor building, the auxiliary building, and the fuel handling building. Detectors inside the containment indicated very high levels of radiation.

Over the period from 4 h 30 min to 7 h into the incident, attempts were made to collapse the steam voids in the two loops by increasing the steam pressure and by sustained HPIS operation. These attempts to reestablish heat removal through the steam generators were unsuccessful and, moreover, involved significant use of the PORV block valve. This course of action was therefore abandoned.

Subsequent calculations of the likely course of events in the reactor over the first 3 h of the incident are illustrated in Figure 5.5 Calculated peak fuel temperatures and calculated core liquid levels (and two-phase mixture levels) are shown. The events referred to in the above description are also indicated. Figure 5.5c shows the temperature calculated at several different levels in the core, level 1 being at the bottom of the core and level 5 near the top.

Phase 5: Extended Depressurization (6–11 h).   Over the next 4 h the operators reduced the pressure in the reactor circuit in an attempt to activate the accumulators and the LPIS components of the ECCS system. This action was initiated at 7 h 38 min by opening the PORV block valve (item 10 in Figure 5.6). At 8 h 41 min, the reactor circuit reached a pressure of 41 bars and the accumulators (item 11, Figure 5.6) were activated. However, only a small amount of water was injected into the vessel.

During the depressurization, a considerable volume of hydrogen was vented from the coolant circuit to the reactor building. At 9 h 50 min a pressure pulse was recorded in the reactor building, and in response the building spray pumps (item 12, Figure 5.6) came on within 6 s and were shut off after 6 min. This pressure pulse was due to ignition of a hydrogen-air mixture in part of the reactor building.

The extended attempt at depressurization was unsuccessful in that the lowest pressure achieved was 30 bars. Nothing that was attempted could drive the pressure lower, and it obstinately remained above the maximum pressure at which the LPIS system of the ECCS could be brought into operation (28 bars).

With the operators unable to further depressurize the reactor circuit, the block valve to the PORV was closed at 11 h 8 min. Over the next 2-h period there was no effective mechanism for removing the decay heat. The block valve was kept closed during this time except for two short periods. Injection via the HPIS was at a low rate and was almost balanced by the outflow through the line to the water purification system; both steam generators were effectively isolated.

Phase 6: Repressurization and Ultimate Establishment of a Stable Cooling Mode (13–16 h).   At 13 h 30 min into the incident, the PORV block valve (item 13 in Figure 5.7) was reclosed, and sustained high-pressure injection via the HPIS was initiated in order to repressurize the circuit and allow the circuit

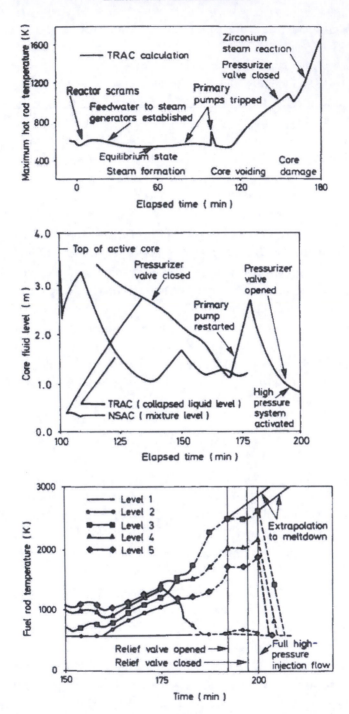

Figure 5.5: TMI-2 results from computer calculations.

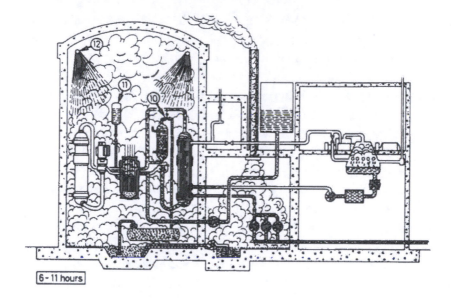

Figure 5.6: TMI-2 phase 5: extended depressurization.

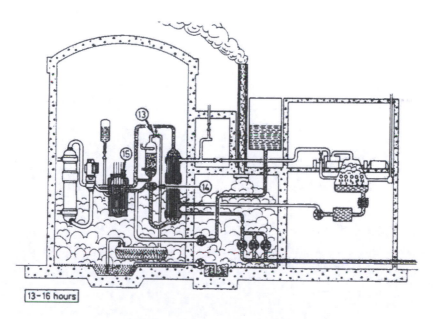

Figure 5.7: TMI-2 repressurization and ultimate establishment of stable cooling mode.

pumps (item 14, Figure 5.7) to be restarted. At 15 h 51 min a circulating pump in loop A was restarted and flow through the steam generators was reestablished, giving a stable heat rejection mode by that means.

Phase 7: Removal of the Hydrogen Bubble (day 1–day 8). As a result of the zirconium-steam reaction, nearly a ton (1000 kg) of hydrogen was produced, and a great deal of this was trapped in the upper region of the reactor pressure vessel, above the core. This "hydrogen bubble" (item 16, Figure 5.8) was eliminated by two methods.

The first method employed the normal purification system used for the primary system. The method worked as follows. The gas in the bubble was being absorbed in the water by the primary system, which was at approximately 70 bars. Some of this water was bled into a "letdown" tank kept at essentially atmospheric pressure, where the absorbed hydrogen gas fizzed out as when a champagne bottle is opened. The gas was passed through a system that delayed its release for 30 days. It was then passed through filters and vented out of the off-gas stack to the atmosphere.

In the second method, heaters in the pressurizer were turned on, forcing the dissolved gas out of the primary system water in the bottom of the pressurizer

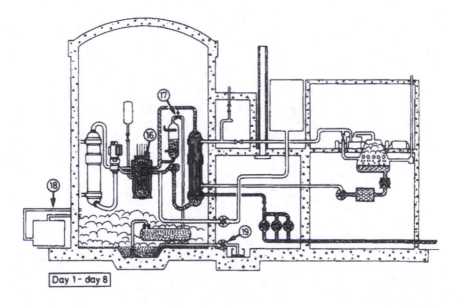

Day 1 - day 8

Figure 5.8: TMI-2 removal of hydrogen bubble.

and into the gas space at the top. The block valve at the top of the pressurizer (item 17 in Figure 5.8) was then opened to permit the gas to escape. The gas bubble was eliminated by these two methods, and on April 28, a month after the accident, cooling by natural circulation was achieved and the reactor coolant pumps were switched off. Switching these pumps off was helpful since the frictional heating of the water by the pumps was at that stage greater than the decay heat being emitted by the reactor core.

Postmortem. Analysis and examination of the damaged core and components have continued in the period since the accident. It is now possible to describe with some confidence the sequence of events that occurred.

Over the first 100 or so minutes with at least some of the main reactor coolant pumps running—albeit circulating a two-phase coolant—the core was adequately cooled (Figure 5.5). Tripping the last coolant pump allowed the steam and water to separate, effectively preventing further circulation through the loops. Gradually, the water in the reactor vessel boiled off exposing fuel 10–15 minutes later (Figure 5.5). However, some decay heat was being removed by steam being released through the open PORV (Section 4.3.2, Figures 4.11 and 4.12). At around 140 minutes the operators closed the PORV block valve, effectively terminating this cooling. The core temperatures rose rapidly above 1800 K. As can be seen from Table 4.2, the cladding would first be oxidized and perforated and, as the temperature increased, a Zircaloy-steam reaction would lead to the formation of hydrogen. Ultimately all the Zircaloy in the affected region would react, and the support given to the fuel pellets would disappear. An estimate of the hydrogen inventory after the accident suggested that about one-third of all the Zircaloy had reacted and almost all the fuel had failed.

The exothermic chemical reaction between Zircaloy and steam increased temperatures still further, taking them above 2400 K. At this temperature Zircaloy is molten and begins to interact with the UO fuel (Figure 5.9*a*). At 174 minutes one of the reactor coolant pumps in loop B was started and operated briefly. The large quantity of water entering the reactor vessel caused the very hot cladding and fuel in the upper part of the core to fragment and collapse (Figure 5.9*b*), leaving an upper crust with a void below. This water achieved some temporary cooling, but the heat-up continued in the lower and central regions of the core. It may be that resolidified material formed a solid crust that acted as a crucible to hold the molten fuel (see Figure 6.1).

At 200 minutes the activation of the HPIS recovered the core and refilled the

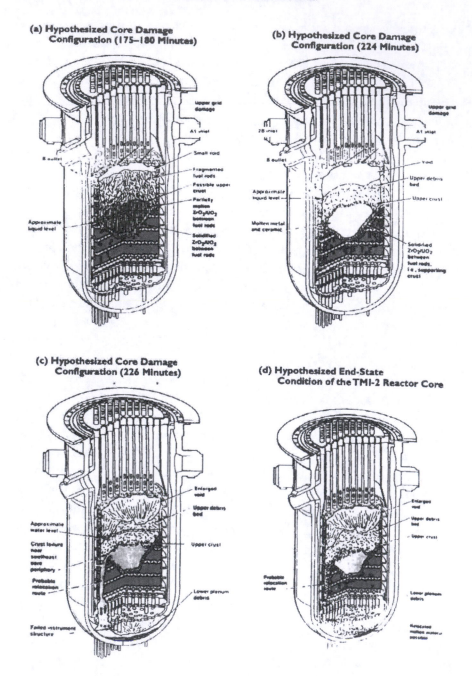

Figure 5.9: Artist's impression of TMI-2 core.

reactor vessel. However, quenching was slow because the water could not reach the seriously damaged areas of the core.

Around 224 minutes it is now known that a major redistribution and reconfiguration of the fuel material took place. The upper crust, left following the formation of the initial void at 174 minutes, now collapsed. Its weight caused molten fuel to be extruded out to one edge of the core where it flowed down over the core support assembly into the bottom head of the reactor vessel. It is estimated that some 20 tons of material ended up in this location (Figures 5.9c and 5.9d). Continued operation of the HPIS finally quenched the core. The slumping of the fuel material increased the resistance of flow through the core, and the flow resistance of the damaged core was estimated at between 200 and 400 times its normal value. At least 70% of the fuel was damaged and 30–40% actually melted.

An international investigation (TMI-VIP) was mounted to examine the extent of the damage to the lower vessel structure and the margin to failure of the reactor pressure vessel. As a result of this analysis it is clear that effective cooling had occurred by penetration of water through cracks in the debris and between the debris and the vessel wall. The molten fuel is also less aggressive to steel than previously feared.

The very high levels of radioactivity in the containment building after the accident were mainly due to the presence of radioactive krypton and xenon. Apart from krypton-85 (which has a 10-year half-life), most of the radioactive isotopes of krypton and xenon are short-lived. With the exception of approximately 10,000 curies of krypton-85, which were vented from the containment about 1 year after the accident, all the radioactive gases escaped in the first few days after the accident, and this led to a measurable increase in activity above the normal background level in the area surrounding the plant. However, very little (only 16 curies) of the iodine released from the fuel escaped from the containment. Evacuation of the area immediately surrounding the site 2 days after the accident involved about 50,000 households. However, exposure of the public to radioactivity was very small indeed, and the consequences in terms of additional cancer deaths are calculated to be undetectable in the surrounding population. Using the estimated total collective dose of 33 man-Sv, it is calculated that there will be less than 1 additional cancer death due to the accident in a total of 325,000 such deaths in the surrounding population over the next 30 years.

A Presidential (Kemeny) Commission investigating the causes of the accident found that operator error was the direct cause. Contributing factors were operator training, control room design, and the attitude toward safety within the U.S. nuclear industry. The Kemeny Commission was also very critical of the Nuclear

Regulatory Commission. The U.S. industry subsequently responded by setting up the Institute of Nuclear Power Operations (INPO) to improve the quality and operational safety of all U.S. nuclear power plants.

The recovery operations for TMI-2 took 10 years and cost about $1 billion. First it was necessary to decontaminate the auxiliary buildings and vent the containment building to allow entry (July 1981). Then the large amounts of contaminated water in the basement of the containment building had to be treated (complete by August 1984). Finally, the reactor vessel had to be opened and defueling undertaken—this took five years (complete by 1990). TMI-2 will be mothballed and dismantled along with TMI-1 around the year 2010.

In terms of the classification of operating states presented in Chapter 4 the incident began as a classical upset transient and then developed (because of the stuck-open PORV) into an emergency condition of the classical small-break type. This should have been easily contained by activation of the engineered safety features, but operator action specifically prevented this from happening. The situation was therefore escalated into an accident beyond the limiting fault condition, that is, beyond the design basis. Nevertheless, the defense-in-depth philosophy of a reactor plant (i.e., the concept of multiple barriers) prevented any significant harm to the public or the operators. Many lessons learned from the TMI accident have been incorporated in newer nuclear plants, albeit at considerable extra cost.

5.2.5   The Ginna Incident

One of the design basis fault conditions for a PWR listed in Chapter 4 is rupture of a steam generator tube. Such an event occurred at the R. E. Ginna PWR station in New York State on January 25, 1982. The Ginna station is based on a two-loop Westinghouse PWR. At 9:25 A.M. the plant was operating at 100% power [490 MW(e)]. Shortly thereafter the primary reactor coolant system pressure dropped significantly, followed by nearly simultaneous activation of the HPIS, reactor trip–turbine trip, and containment isolation. The pressurizer went almost empty. This is the behavior expected when a steam generator tube bursts (ruptures), allowing primary circuit water to pass into the (lower-pressure) secondary side of the steam generators. Following standard procedures for responding when it is suspected that a steam generator tube has ruptured, the operators tripped the main coolant pumps and closed the main steam isolation valves on the suspect steam generator.

The operators opened the PORV connected to the pressurizer in order to equalize the primary and secondary circuit pressures quickly and stem the leak.

This action allowed reactor coolant to drain into the pressurizer relief tank. However, when this operation was completed and an operator tried to close the PORV again, it failed to close (as at TMI-2), requiring the operator to shut the block valve and thus isolate the flow, which he did promptly. The depressurization resulting from the opening of the PORV caused flashing in the primary circuit, pushing water into the pressurizer and causing a void to form in the top part (upper head) of the reactor vessel. This situation was recognized and rectified by starting a main circulating pump some 2 h after the start of the incident. No excessive fuel temperatures were noted.

The operation of the PORV caused the pressurizer relief rupture disk to blow and some 5,000–10,000 gallons of water drained into the sump of the containment building, which had been isolated by this stage. During this time, the damaged steam generator was isolated on the secondary side and the pressure in this steam generator went up to the point where the secondary relief valve lifted. This resulted in a minor radioactive release to atmosphere, mainly of krypton and xenon.

The plant was subsequently cooled down, first by using the undamaged steam generator to remove the residual heat and then, after about 24 h, by the low-pressure residual heat removal system.

Subsequent inspection of the damaged steam generator showed that a loose pie-shaped metal object, weighing about 2 lb, was present in the steam generator. This object had vibrated and severely damaged a number of steam generator tubes, causing one of them to rupture and leading to the events described above. The object appears to have been present in the steam generator for a number of years, having been introduced inadvertently during earlier maintenance work. Flow through the damaged tubes was blocked by plugging them, and the unit has been returned to power.

Unlike the accident at Three Mile Island, the operator response at Ginna was good, although somewhat delayed compared with the operating guidelines for this type of incident. Although the Ginna incident has received the most publicity, it is noteworthy that steam generator tube ruptures had occurred previously, one example being an incident in PWR Unit No. 2 at Prairie Island, Oregon, on October 2, 1979.

5.2.6 The Mihama-2 Incident

The Mihama-2 power station in Japan is equipped with a 500-MW(e) two-loop pressurized water reactor. On February 9, 1991, this plant also experienced a steam generator tube rupture that allowed high-pressure water from the reactor circuit to flow into the (lower-pressure) secondary circuit formed by the steam generator shell, the turbine, and the condensers.

At 12.24 h, with the reactor at full power, increasing radioactivity was signaled in the blowdown line of one of the plant's steam generators. Further indications from the water in the steam generators and in the air extract from the condenser suggested an initially minor leakage of primary coolant water from a damaged tube in this steam generator.

Around 12.45, both the pressure in the primary circuit and the water level in the pressurizer started to decrease despite the activation of the pumps charging water into the coolant circuit. Three minutes later, a reduction of reactor power was initiated. The isolating valves on the steam line from the affected steam generator were activated but failed to close completely. Two minutes later, the reactor, turbine, and generator were automatically shut down. Almost immediately, reducing pressure and level in the pressurizer signaled an emergency core cooling system (ECCS) water injection. At the same time, the reactor containment was automatically isolated. Primary loop pressure and water levels continued to reduce rapidly.

At 13.52, the feed flow to the damaged steam generator was stopped and the unit was isolated. Ten minutes later, the relief valve on the steam line from the remaining undamaged steam generator (B) was lifted to allow decay heat removal. Attempts were also made to equalize the pressures on the primary and secondary sides by opening the pressurizer relief valves. However, this operation was not successful, and depressurization of the primary circuit was therefore undertaken via the alternative of the pressurizer spray system.

By 14.34, the HPIS pumps were stopped and 12 minutes later the primary and secondary circuit pressures were equalized, terminating the release. The reactor reached a safe, "cold standby condition" at 02.30, February 10, 1991.

Analysis after the event suggested that 55 tons of coolant passed from the primary to the secondary systems and 1.3 tons of radioactive steam bypassed the main steam isolating valve and were released to the turbine hall.

After the accident, a camera was lowered into the damaged steam generator and located the fractured tube. The tube was removed for inspection. It had ruptured close to the top (sixth) tube support plate (Figure 5.10). There was evidence of fatigue failure due to vibration of the tube. Corrosion debris was also found between the tube and the support plate.

This type of recirculating steam generator is equipped with V-shaped antivibra-

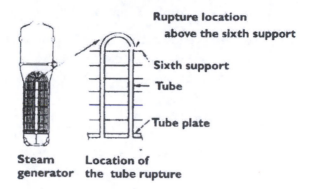

Figure 5.10: Location of the ruptured steam generator tube.

tion bars to prevent fluid-induced tube vibration in the return bend region. However, fiber optic observations showed that these antivibration bars had not been
correctly installed during construction and did not support the damaged tube
(which was also restrained by the debris at the tube support plate; Figure 5.11).

The valve malfunction—the improper seating of the main steam isolating
valve and the inoperative pressurizer valves—was also investigated.

The Mihama-2 incident was a classic steam generator tube rupture–small loss
of coolant event, and the releases of radioactive rare gases and iodine were
small. The plant was out of operation for a considerable period of time while its
steam generators were replaced.

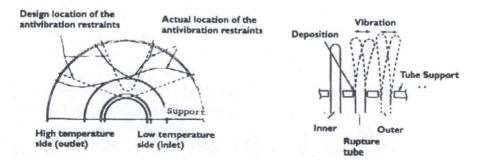

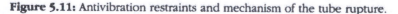

Figure 5.11: Antivibration restraints and mechanism of the tube rupture.

5.2.7 The Serious Accident at Chernobyl

On April 26, 1986, the worst accident in the history of commercial nuclear power generation occurred at the Chernobyl Nuclear Power Station some 60 miles north of Kiev in the Ukraine, on the Pripyat River not far from the town of Pripyat (poplation then 49,000). The site at that time had four 1000-MW(e) RBMK reactors operational and two more under construction. The four reactors were built in pairs, sharing common buildings and services. Construction of Units 3 and 4 started in 1975–76; Unit 4 became operational during 1984. The main elements of the reactor are described in Section 2.4.6.

The Experiment. Ironically the immediate cause of the accident that wrecked the No. 4 Unit was an experiment designed to improve the safety of the plant. The objective of this experiment was to see whether the mechanical inertia in a turbine generator isolated from both its steam supply and the grid could be used to supply electricity via the station distribution system to important station auxiliary loads (including the emergency cooling pumps) for a short period (40–50 seconds). In essence, this was an attempt to use the turbine generator as a mechanical flywheel coupled to the pumps electrically.

A turbine generator unloaded normally would take about 15 minutes to come to rest from 3000 rpm, but when coupled to the pump motors might provide a few tens of seconds' supply. Even so, given the rapid coast-down of the main circulating pumps without this provision and the long time required to shut down the reactor and start the auxiliary diesel generators and diesel, this "flywheel" effect could have provided a valuable margin in the safety case. In the experiment, to simulate the load from the ECCS, the generator was coupled to four of the main circulating pumps (each rated at 5.5 MW) and the feedwater pumps.

The experiment had been attempted twice before, in 1982 and 1984. On the latter attempt, following isolation of the generator from the grid, the voltage level in the unit system fell rapidly and the operators were unable to arrest the drop by manual control of the voltage regulator. The fall in voltage resulted in the pump motors slowing down much faster than the generator.

For the fateful experiment on April 26 an automatic voltage regulator, acting on the generator excitation current, had been fitted that maintained the voltage level in the unit system so that the pump motors ran down in step with the main generator at synchronous speed, drawing on the stored kinetic energy of the turbine generator.

The planned experimental initial conditions required the reactor to be at about 25% full power with one of its turbine generators shut down and the other supplying the grid, four main circulating pumps, and two feed pumps. The remaining auxiliary plant was fed from the grid.

The experiments had been badly planned, the safety case was inadequate and had not been properly reviewed, and as we shall see in the following sections, the operators failed to achieve the chosen plant conditions, departed from the laid-down procedures, and violated several operating rules.

Status of the Plant before the Accident.   On April 25, 1986, all four units at Chernobyl were operating. The No. 4 unit was due to be shut down for maintenance work. A total of 1,659 channels were loaded with fuel, most of it (75%) from the initial fuel charge, having been utilized to an extent ("burn-up") of 12–15 MW, day kg. What follows is an abbreviated and simplified account of the sequence of events that took place. For ease of description the accident is divided into a series of logical phases. Diagrams illustrate the condition at each phase.

Phase 1: Prelude [01.00–23.10 h, April 25 (Figure 5.12)].   The reactor was at nominal full power conditions [1000 MW(e), ca. 3000 MW(t)]. The operators started to reduce power at 0.100 h, on April 25, and about 12 hours later, at 13.05 h, with the reactor at 1600 MW(t), turbo generator No. 7 was disconnected from the grid. Four of the main circulating pumps and two of the feedwater pumps were connected to turbo generator No. 8 in preparation for the test.

At 14.00 h, the emergency core cooling system was disconnected from the primary circuit. This was in accordance with the experimental plan (presumably because it was anticipated it would be spuriously initiated by the expected low level in the steam drum during the experiment).

However, the grid controller requested the unit to continue supplying to the grid until 23.10 h. Operation with the emergency core cooling system disengaged was a violation of the operating rules (*violation 1—one of many to come*), but it does not appear to have had any significance in the accident sequence. However, disabling of the reactor protection system seems to have been regarded rather lightly both in the operating procedures and by the operators themselves.

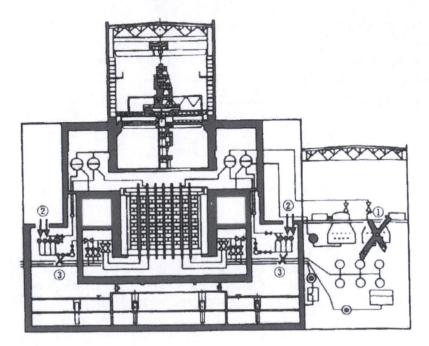

Figure 5.12: Phase 1: prelude (01.00–23.10 h, April 25, 1986) (X indicates components not in operation at time of accident).

Phase 2: Preparations for the Experiment (23.10 h, April 25, to 01.00 h, April 26).   At 23.10 h, the operators started to reduce power to obtain the test condition of 700–1000 MW(t). The local automatic control (LAR) system, which operated 12 control rods, was disengaged at 00.28 h on April 26. Here the operator made a major error (*violation 2*) in failing to reset the set point of the automatic regulation (AR) system and was then unable to control the reactor power with a combination of the manual and overall automatic control (AR3), the latter using only four control rods. The result was that the reactor power dipped to below 30 MW(t).

The first reduction from 100% power nearly 24 h earlier had initiated a xenon poisoning transient. The fission product Xe-135 is of considerable importance in thermal reactors because it has a very high neutron capture cross section. Only a small proportion of Xe-135 is formed directly by fission; most comes from the radioactive decay of I-135 (half-life 6.7h). The xenon is removed partially by decay (half-life 9.2h) and partly by its capture of neutrons. About 2% of all neutrons are captured by Xe-135, so it is an important item in the overall neutron balance (see Section 2.2). The balance of formation of xenon and its destruction are such that a fall in reactor power (and thus of neutron flux) leads to a rise in xenon concentration.

Figure 5.13*a* shows the reactor power-time history together with (Figure 5.13*b*) the poisoning effect of the xenon present. It will be seen that the peak in the transient (at about 12–14 h after the initital decrease in power) had passed but that the uncontrolled drop in power to around 30 MW(t) had induced a sharp increase in the xenon poisoning by the time the experiment started. Because of the sharply increasing xenon the operator had considerable difficulty in raising reactor power with the small operating reactivity margin he had available. Finally, at 01.00 h on April 26, the power was stabilized at 200 MW(t)—well below the power level proposed for the experiment.

Phase 3: The Experiment [01.00–23.40 h, April 26 (Figure 5.14)]. At 01.03 and 01.07 h, respectively, the operators started the main standby circulating pumps (see 4 in Figure 5.14), one on each main loop, so that at the end of the experiment, in which four pumps were to operate "tied" to the No. 8 tur-

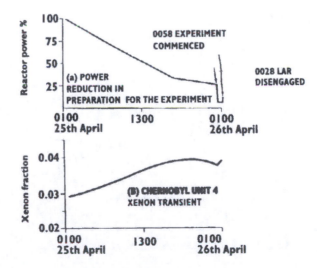

Figure 5.13: Chernobyl Unit 4 xenon transient.

bine generator, four pumps would remain coupled to the grid to provide reliable cooling of the core.

The reactor power was lower than intended; so too were the steam voidage in the channel and the pressure drop along the fuel. As a result the coolant flow rate was higher than anticipated with all eight pumps operating. Such an operating mode was normally prohibited because of the possibility of single-pump trip leading to cavitation and vibration of the main feed piping (*violation 3*).

Because the reactor power was only 7% of full power and the coolant flow rate through the core was 115–120% of normal, the *enthalpy* rise across the core was only 6% of nominal, or equivalent to just 4°C. Thus although the entire primary coolant system was only slightly subcooled and still very close to boiling, there was very little steam being generated in the core.

Under these conditions the coolant voidage would have been much reduced. The water was absorbing more neutrons, so the control rods were correspondingly further withdrawn. The decrease in steam generation resulted in a drop in steam pressure and disturbances to other reactor parameters. The oper-

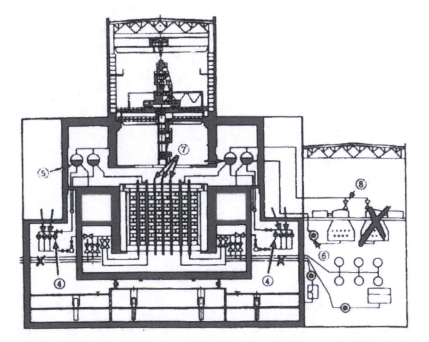

Figure 5.14: Phase 3: the equipment (01.00–01.23 h, April 26, 1986).

ators tried to control both the steam pressure and the drum level manually but were unable to hold these parameters above the normal "trip" point settings (5 in Figure 5.14). To avoid the reactor's tripping, the operators overrode the trip signals with respect to these variables (*violation 4*).

At 01.09 h (4 minutes before the initiation of the test), the operator opened the main feed valve (6 in Figure 5.14) to increase the water level in the steam drum. With the feedwater flow increased by a factor of 3, the desired water level was reached 30 seconds later. However, the operator continued to feed the drum. As the cold water from the drum passed into the core, the steam generation rate fell noticeably, resulting in an even further reduction in steam voidage. To compensate, all the 12 automatic control rods moved upward to a "fully withdrawn" position (7 in Figure 5.14).

To maintain reactor power at 200 MW(t) the operator had also to move a number of manual control rods up. This allowed one group of automatic control rods to reenter the core by 1.8 m.

The cool feedwater and the decrease in steam generation led to a small fall in pressure. At 01.19.58 h, a steam bypass line to the condenser was closed, but the steam pressure continued to fall (by 5 bars) over the next few minutes.

At 01.21.50 h, the operator sharply reduced the feedwater flow rate, which resulted in an increase of water temperature passing to the inlet water with a delay of the transit time (20 s) from the steam drums to the reactor inlet. The automatic control rods started to lower into the core to counter the effect of the increased voidage.

At 01.22.30 h, the operator looked at the printout of the reactor parameters, especially the residual reactivity margin left in the control rods. Over this period the control rods remained substantially withdrawn.

A "safe" operating level was set to ensure that the control rods "dipping" into the core were effective when they moved. The operator noticed that the reactivity margin was at a value (less than 15 rods inserted into the core) that required him to trip the reactor. The test was, however, continued in violation of this operating restriction (*violation 5*).

Calculations have shown that the number of control rods in the core at this stage was 6 to 8—less than half the design "safe" minimum and a quarter of the minimum number of 30 inserted rods given in the operating instructions (related to a negative reactivity insertion rate of 0.5–0.7% / s).

It should be observed that measurements from in-core flux monitors showed the neutron flux profiles to be normal in the radial plane but doubly peaked in the axial direction with the higher peak in the upper region of the core. This was caused by high xenon levels in the central part of the reactor, coupled with steam generation in the upper parts of the core.

At 01.23.04 h, the experiment was initiated and the main steamline valves to turbine generator No.8 were closed (8 in Figure 5.14). The protection provided to trip the reactor when both turbine generators were tripped had been disengaged to allow the reactor to continue to operate. However, this was not part of the original plan for the experiment and was done apparently to enable the test to be repeated if the first test was unsuccessful. Needless to say this was a further violation of the operating procedures (*violation 6*). The operation of the reactor after the start of the experiment was not required.

The No. 8 turbine generator together with the four main circulating pumps (see 2 in Figure 5.14) and two feedwater pumps (6 in Figure 5.14) started to run down. With the closure of the main steam and bypass valves the steam pressure rose slightly and the steam generation in the core correspondingly decreased slightly (01.23.10 h). However, the main coolant flow and the feedwater flow reduced, causing an increase in both water inlet temperature and steam generation. An increase in reactor power was noted at 01.23.31 h. An attempt was made to compensate with the 12 automatic control rods, but this was ineffective.

A power excursion was experienced, and at 01.23.40 h, the shift manager attempted a manual "scram" of the reactor. All the control rods and emergency rods began motoring into the core. However, the rods could not be fully inserted. Because the rods were in a nearly withdrawn position, a delay of about 10 s occurred before the reactor power could have been reduced. Indeed, the very act of driving in the "overdrawn" control rods may have contributed to the initiating event for what followed. The control rod "followers" (see Figure 2.14) displaced the neutron-absorbing water on reinsertion to start the power excursion.

In this time a prompt critical power excursion driven by the increased steam generation in the core (due to the pump rundown) and the strong positive void coefficient led to severe fuel damage and fuel channel disruption. After 3 s the reactor power had reached 530 MW(t) and continued to increase exponentially to much higher levels. Only the negative fuel temperature coefficient (Doppler effect) was acting to reduce the neutron population over this period. The specific energy deposited in the fuel was estimated to be greater than 1.2 MJ/kg. There were two excursions in power. It has been suggested that the second power peak was from additional voiding caused in turn by the rupture of the pressure boundary during the first excursion.

The condition of *prompt criticality* (see Section 2.3) is believed to be what occurred in the last stages of the accident at Chernobyl. Complete voiding of the RBMK core would have produced about a 3% increase in k, greater than the delayed neutron fraction.

At 01.24 h, witnesses heard two explosions, one after the other. Molten and burning fragments flew up from the Unit No. 4 plant and some fell on the roof of the turbine generator building, starting a fire.

Phase 4: Explosion and Fire [01.23.40–5.00 h, April 26 (Figure 5.15)].
The precise sequence of events following the reactivity insertion will probably never be known, but based on analysis, actual observations, and previous experimental work a plausible picture can be put together.

One particularly relevant experiment is that undertaken in 1979 at the Power Burst Facility (PBF), Idaho Falls, as part of the Thermal Fuels Behavior Program for the USNRC. A single unirradiated UO_2 fuel rod, operating under conditions representative of hot start-up of a boiling-water reactor (i.e., very similar to the

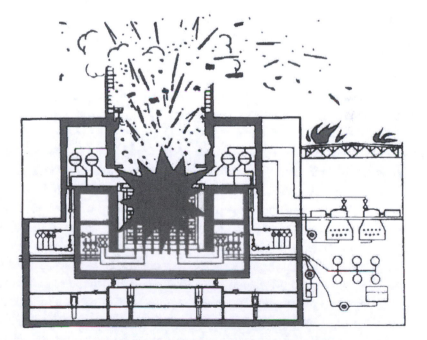

Figure 5.15: Phase 4: explosion and fire (01.23.40–05.00, April 26).

conditions at Chernobyl) was subjected to a power burst, resulting in a total energy deposition of 1.55 MJ/kg UO_2 (cf. Chernobyl about 1.2 MJ/kg UO_2).

Extensive amounts of molten fuel debris were expelled into the flow channel and against the pressure tube wall. A pressure pulse of 350 bars, suggesting an energetic molten fuel-coolant interaction, was observed. Following the modeling of the accident, it would appear that in the case of the Chernobyl transient the energy deposited in the fuel from the power transients probably resulted in fuel melting or fuel fragmentation and dispersion. The fuel cladding initially remained intact until voiding in the channel-induced "dryout," after which the clad temperature increased at 250°C/s. The subsequent explosive formation of steam caused a sharp increase in the pressure within the fuel channel sufficient to increase the steam drum pressure at ~10 bars/s and to stop or even reverse the primary coolant flow. This is known because the check valves downstream of the pumps closed at 01.23.45 h. This further voiding of the fuel channels resulted in a second, larger power surge to about 440 times full power.

Fuel ejected from the fuel pins under the driving force of fission gas pressure impinged on the pressure tubes, causing failure and releasing steam into the graphite moderator space. With the pressure relieved at 01.23.47 h, water rushed back into the fuel channels to interact with the fuel being ejected from the fuel pins. A conservative estimate of the total thermal energy deposited in the fuel is 50–100 GJ. Assuming a 1% efficiency for the conversion to mechanical energy in an energetic fuel coolant interaction (FCI), a conservative explosive energy of 0.5–1 GJ is estimated. This is broadly equivalent to 100–200 kg of TNT (but, in the case of the explosive, detonation is much more rapid than in the FCI). The conditions were also appropriate for other chemical reactions including molten zirconium–steam and hot graphite–steam reactions. At 0.23.48 h, two explosions were noted in succession; the first could have resulted from the fuel-coolant interaction and the second from hot hydrogen and carbon monoxide mixing with air and exploding as the containment of the reactor vault failed. These detonations, together with the buildup of steam pressure, blew the 1000-ton top shield off and rotated it through 90° (Figure 5.16). It also broke all the pressure tubes and lifted some of the control rods. Some of the graphite blocks from the reflector were ejected, the charge face was destroyed, and damage was done to the charge hall and some of the structural parts of the building. Fragments of core materials fell onto the roofs of the reactor and turbine buildings. The refueling machine that stood on the charge face "leapt up and down," causing further pipework failures. Over 30 fires were started in var-

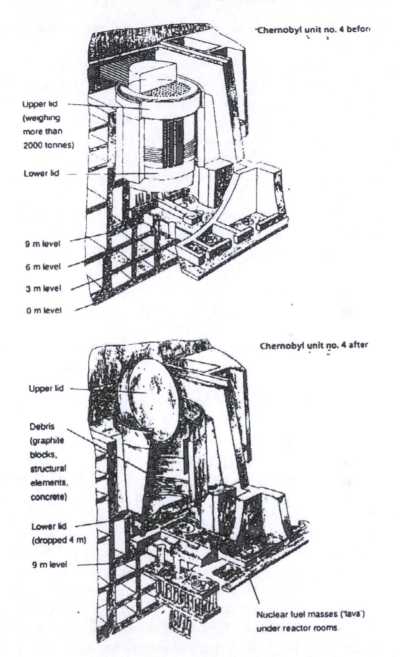

Chernobyl unit no. 4 before

Upper lid
(weighing
more than
2000 tonnes)

Lower lid

9 m level

6 m level

3 m level

0 m level

Chernobyl unit no. 4 after

Upper lid

Debris
(graphite
blocks,
structural
elements,
concrete)

Lower lid
(dropped 4 m)

9 m level

Nuclear fuel masses ('lava')
under reactor rooms.

Figure 5.16: Chernobyl Unit 4 before and after the accident.

ious areas due to ruptured fuel lines, damaged cables, and thermal radiation from the exposed core.

By 01.30 h, the firefighters on duty had been called out and were reinforced with firefighting units from Pripyat and Chernobyl. Graphic accounts have been given of the extreme heroism of these firefighters, many of whom have since perished as a result of their exposure to lethal doses of radiation. By 05.00 h, the fires on the reactor and turbine buildings had been extinguished. Amazingly, the three other units at the station continued to operate. The No. 3 Unit, which was adjacent to the damaged unit, was not shut down until 05.00 h. The other two units continued to operate until the early hours of the following morning, some 24 h after the accident. Fuel temperatures, initially high due to the energy deposited in the transient, fell as the heat was transmitted to the graphite and other reactor components.

Phase 5: The Aftermath (05.00 h, April 26 to May 6). With the reactor core badly damaged and the cooling system not functional, the Soviet engineers started to consider how to fight the graphite fire and how to reduce core temperatures, deal with the decay heat, and limit fission product release. They initially tried to cool the core by the use of emergency and auxiliary feedwater pumps to provide water to the core. This was unsuccessful. Given the continuing graphite fire and ongoing significant release of fission products, the decision was taken to cover the reactor vaults with boron compounds, dolomite, sand, clay, and lead. The boron was to stop any recriticality; the dolomite gave off CO_2 as it heated up (which reduced the access of oxygen to the graphite fire); the lead absorbed heat, melted into gaps, and acted as shielding; while the sand acted as an efficient filter.

Over the period April 27–May 10, over 5000 tons of materials were dropped by military helicopters. The reactor core was thus covered by a loose mass that effectively filtered the fine aerosol fission products. Around May 1, some 6 days after the accident, fuel temperatures started to increase due to fission product decay heating and graphite combustion. To reduce temperatures, compressed nitrogen was fed into the space beneath the reactor vault. Fuel temperatures peaked about May 4–5 at around 2000°C and then began to drop. It is believed that about 10% of the core graphite was consumed during this period. By May 6, the discharge of fission products had virtually ceased, having decreased by a factor of several hundred.

Phase 6: Stabilization and Entombment [from May 6 (Figure 5.17)].
From early May the situation at the damaged reactor improved. Monitoring devices to measure temperatures and air speed were lowered into the debris. The exact disposition of the fuel in the damaged reactor is not known. By May 6 at least 60–80% of the fuel had been released from the reactor vessel itself. About 130 tons of the molten radioactive material from the core formed into a "lava" most of which found its way to the lower parts of the reactor building.

From May 6, temperature conditions in the reactor vault were stable at several hundred degrees centigrade but falling at 0.5°C/day, fission product releases were down to tens of curies/day, and radiation levels in the areas immediately adjacent to the reactor were at levels of single sieverts per hour. Further fires broke out on May 23 in the plant areas above the damaged reactor. Although these were in high-radiation zones, they were successfully dealt with.

The worry was that the molten debris would melt through the last 50 cm of a 2-m-thick concrete slab at the 9-m level. A flat concrete slab incorporating a heat exchanger was designed and installed in the area beneath the reactor vaults by the end of June. A decision was taken to entomb the critically damaged unit in protective concrete walls 1 m thick. This included a perimeter wall enclosing the turbine and reactor blocks as well as internal and dividing walls between Units 3 and 4 and a protective cover over the turbine and reactor blocks. An internal recirculating ventilation-cooling system was installed, and the entombed reactor was maintained at reduced pressure (in respect of atmospheric pressure) and the exhausted air discharged through filters and a stack. This work was completed by early autumn of 1986. However, the "sarcophagus," as it is known, did not remain leak-tight for long and there continue to be concerns about its integrity and the up-ended top shield–reactor roof.

Consequential Events and Core Damage.   The reactor core was very severely damaged by the explosion, which also caused structural damage to the reactor building. A considerable discharge of fission products took place (Figure 5.18), and it is estimated that excluding the noble gases, 70 megacuries (when related to the time of the reactor shutdown, $\sim 2.6 \times 10^{18}$ Bq) were released in essentially two periods: the initial explosion and early stages of the graphite fire (April 26–27) and the later heat-up transient (May 2–5). This total release corresponds to 3–3.5% of the total fission product inventory—some 6–7 tons of material.

Of this, some 0.3–0.5% (0.6–1 ton) is estimated to have remained on the site,

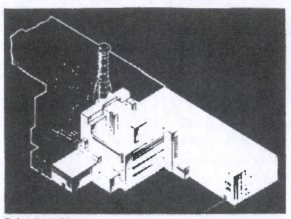

Before the incident . . .

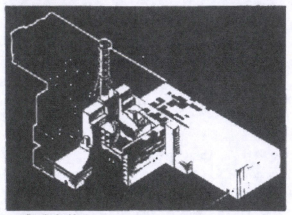

. . . after the incident

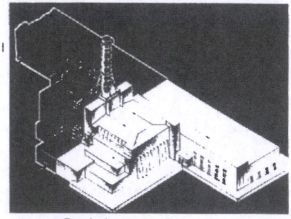

. . . after entombment

Figure 5.17: Chernobyl Unit No. 4 before and after entombment.

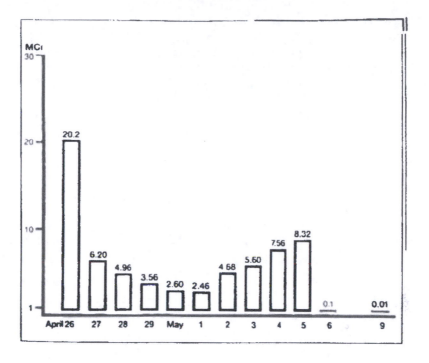

Figure 5. 18: Daily radioactive releases into the atmosphere from the accident (without radioactive noble gases) (1 MCi = 37 x 10^{15} Bq = 37E Bq).

with 1.5–2.0% (3–5 tons) being deposited within 20 km and 1.0–1.5% (2–3 tons) being transported to greater distances. Particle sizes of the released material ranged from below 1 micron to 10s of microns.

Table 5.1 shows the estimated fractional releases of fission products during the accident. Most of the gaseous fission products (Xe, Kr) were released together with significant amounts of I^{131} and Cs^{137} as well as smaller amounts of fuel aerosol material produced by corrosion of UO_2 exposed in the mechanical and thermal disruption of the reactor core. Table 5.1 shows the corresponding releases for the accident at Three Mile Island, both for the release from the reactor core and the release to the environment.

It will be seen that the extent of fuel damage and fission product release from the core in the two accidents is very comparable. However, the effectiveness of the containment and the ECCS in preventing any significant release to the environment in the case of TMI-2 is dramatically clear.

At Chernobyl two operators were killed by falling debris and burns during

Table 5.1 • Three Mile Island and Chernobyl Releases Compared

	TMI-2: Outside the Core	TMI-2: To Environment	Chernobyl: To Environment
Noble gases (Xe, Kr)	48%	1%	76
I	25%	3×10%	15–20%
Cs	53%	not detected	10–13%
Ru	0.5%	not detected	2.9%
Ce (group)	NIL	not detected	2.3–2.8%

the first few hours after the accident. Up to the end of August 1986, a further 29 people, all involved in firefighting or other accident recovery methods, had died of massive doses of radiation. About 200 staff received high radiation doses and burns.

A 30-km radius control zone was established around the Chernobyl site. Pripyat, Chernobyl, and other population centers were evacuated from April 27 onward: in all, about 135,000 people plus several thousand livestock. In 1990, 50,000 more people were evacuated, and further evacuations have occurred since, although the average doses delivered by the environment directly are now low.

A massive effort was undertaken to decontaminate the Chernobyl site (to permit entombment of the No. 4 Unit and the return to operation of the other undamaged units) and the surrounding 30-km zone. Special measures were devised to protect ground and surface water from contamination by way of curtain walls between the reactor site and the Pripyat River. In all, about 650,000 persons involved in the cleanup of the plant site and the 30-km zone were exposed to radiation.

Very extensive areas of the former Soviet Union and beyond its frontiers were affected by fallout. The plume from the initial fission product releases reached a height of 1200 m. The cloud was generated over several days (Figures 5.19 and 5.20). Initially the cloud traveled northwest, missing Pripyat, across the Soviet Union and northeast Poland to Scandinavia. Some days later it changed direction and swung southward across Poland and Central Europe. Figure 5.19 shows the likely trajectories of materials reduced from Chernobyl on April 26.

Heavy rain on April 30 and May 1 led to wet deposition of radioactivity

Figure 5.20: Caesium-137 (Bq m⁻²) in vegetation in the United Kingdom. (From the Institute of Terrestrial Ecology.)

there were also no indications of an increase in leukemias and cancers. There were, however, significant non-radiation-related health disorders in the population surrounding Chernobyl. Nine years after the accident, many of the expected health effects had not become apparent because of the latency period for some radiation-induced cancers. So the health effects can be summarized as:

- Acute radiation sickness and burns from β radioactivity to some 200 people causing 28 deaths.
- Childhood thyroid cancer in children living in and around Belarus and the northern district of Ukraine. So far nearly 500 cases of childhood thyroid cancer (associated with the uptake of I^{131}) have been detected in a population of 3 million children at risk.
- Nonradiological effects from stress-related conditions in a population of 10 million living in the most affected regions.

On the basis of past experience, some further health effects may be observed in across France, Switzerland, southern Germany, and Czechoslovakia. On Friday, May 2, the cloud reached Britain. While the cloud cleared southern and eastern Britain on May 2 and 3, heavy rain occurred in North Wales, Cumbria, and Scot- land, causing relatively high levels of Cs^{137} activity (Figure 5.20). From May 3, the cloud passed more to the south, over Yugoslavia, Italy, and Greece.

An assessment of the implications of this spread of radioactivity over Europe can only be approximate. The United Kingdom's National Radiological Protec- tion Board has estimated a collective effective dose integrated over all time of 80,000 man Sv. Current estimates indicate perhaps 30,000 fatal cancers resulting over the next 40 years in the affected parts of Russia and Western Europe. This value needs to be compared with over 30 million cancer deaths expected in the same population over the same time period.

In 1991 the International Atomic Energy Agency issued the results of a major study, the International Chernobyl Project, looking at the health effects of the accident. It involved about 200 independent experts from 22 countries and seven international organizations. It concluded at that stage that there were no health disorders that could be directly attributed to radiation exposure and

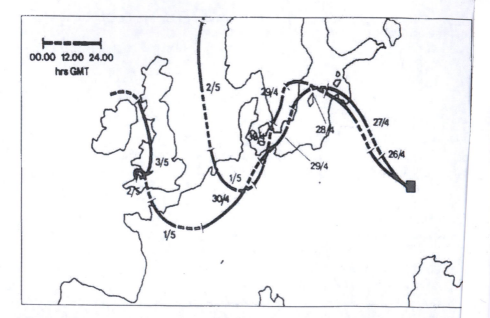

Figure 5.19: MESOS trajectories orginating from Chernobyl at 09.00 h, 12.00 h, and 15.00 h. GMT on April 26, 1986 (ApSimon, et al., 1986).

the 100-km-radius regions around the plant, particularly in relation to breast cancer and skin and lung cancers.

In Britain, restrictions were imposed on the movement and slaughter of sheep and lambs grazing on caesium-contaminated grass in North Wales, Cumbria, and Scotland; originally about 4 million sheep out of a national flock of 25 million were subjected to controls. By March 1988, the number had been reduced to 300,000, but some controls were still in place as recently as 1995. The extra radiation dose received due to inhalation from deposited activity or through the food chain is expected to be on average about 3.5% above the normal annual dose due to natural background radiation (70 micro Sv. in about 2000 micro Sv.). This increase, however, varies from 10% in the north and west to just 1% in the south of the country.

Causes of the Accident.   Given the magnitude and severity of the accident and the fact that other reactors of this type were still in operation, the (then) Soviet Union established a Government Commission to study the causes of the accident. In its report to the IAEA Chernobyl Post Accident Review Conference in August 1986, the Soviet delegation acknowledged that a number of factors had contributed to the accident. Underlying the specific design and operational aspects of the accident were the institutional and organizational shortcomings of the Soviet nuclear industry. Since the accident, many analyses have been undertaken and published. The general conclusion from these analyses of the Chernobyl accident is that no *new* reactor safety issues have been identified.

One unusual, perhaps remarkable, feature of the Chernobyl accident is that failure of equipment played no part in the events leading up to the explosion. Likewise, only one of the actions taken by the operators—*violation 2,* failing to reset the set point of the automatic regulation system at 00.28 h on April 26— can be considered a mistake. All the other violations of the operating rules were deliberate with the specific objective of completing the voltage regulation experiment.

Design Shortcomings. First, the concept and design of the reactor itself was the major contributory factor. While the RBMK reactor has some inherent features that made it quite attractive (including the lack of a thick-walled pressure vessel, the absence of steam generators, the capability to replace fuel on load, and ease of construction on remote sites), it also has features that were shortcomings:

1. *Positive power coefficient at low power levels.* The power coefficient and de-

sign of a reactor dictates its behavior and stability. If the power coefficient is negative, any power rise will be self-limiting; if positive, the converse. The power coefficient is made up of a number of individual components, but in the case of the RBMK, two components are dominant: the negative effect of fuel temperature (Doppler) increases and the positive effect of an increase of steam voidage in the core. At power levels below 20%, the positive void coefficient becomes much stronger than the negative fuel temperature coefficient. As a result the power coefficient is overall positive and the reactor unstable.

2. *Slow shutdown system.* The reactor control and protection system was too slow and inadequate in design. The shutdown system was dependent for its effectiveness on appropriate operation of the reactor control system, which was complex and largely manual. Because computers were rudimentary and unreliable when the RBMK reactor was originally conceived, the designers assumed that human operators would be more reliable. They failed to see the need for engineered safeguard features to counteract the operator's driving the reactor into extreme situations for which the slow shutdown system would be ineffective.

3. *"Positive scram."* Associated with the poor design of the protection system is the design feature that with the control rods fully withdrawn, the initial effect of insertion is to increase reactivity in the lower parts of the core, due to the displacement of water by the graphite followers. Normally, the entry of the boron carbide absorbers would reduce reactivity at the top of the core and overwhelm this increase. However, in the specific sequence of April 26, 1986, because of the double-peaked axial flux profile resulting from the xenon transient, this was not the case. The converse happened: entry of the control rods initially produced either a neutral or even a slight increase in reactivity—"positive scram."

4. *Design of containment.* This was inadequate to cope with this extreme accident. The RBMK reactors do not have a common containment to cover both the reactor and primary circuit.

These unfavorable features, either individually or in combination, are inconsistent with Western safety design principles and would not have been licensed or built in the West.

Operator Violations. Clearly the operators had violated a number of operating regulations vital for the safe operation of the plant, but these only magnified the design shortcomings, particularly at low power. The most serious violations have been highlighted earlier.

It is appropriate to ask why the operators seemed prepared to violate so many operating rules. The explanation seems to be that no serious consideration had been given to the safety aspects of the experiment. The Soviet Government Commission report states: "Because the question of safety in these experiments had not received the necessary attention, the staff involved were not adequately prepared for the tests and were not aware of the dangers." It seems the experiment was regarded as simply another electrical test. At the same time, operators reportedly felt they were under extreme pressure to complete the planned experiment that night since they knew it could be a full year before they had another chance. Other factors could also have influenced the operators to cut corners: The Chernobyl station was "top of the league" for availability, the experiment was delayed (by grid control) and came at the end of a working week early in the morning, and it was the eve of the May Day holiday.

Institutional and Organizational Shortcomings. In addition, shortfalls in managing the safe operation of the power plant were a major contributory cause, and a number of local and central government staff were removed from their positions and convicted of negligence. A separate Ministry of Nuclear Energy was set up alongside the Ministry of Power and Electrification. Professor Legasov, head of the Soviet delegation to the August 1986 IAEA Conference, in his memoirs (he died on April 27, 1988, the second anniversary of the accident) noted the many instances when expediency overcame quality—poor construction, defects in design and manufacture not rectified, etc. Most of all he was critical of the management of safety in the Soviet Union. "The level of preparation of serious documents for a nuclear power plant was such that someone could cross out something and the operator could interpret, correctly or incorrectly, what was crossed out and perform arbitrary operations." This has been described succinctly as a lack of a *safety culture.*

The Remedies. Russia and Ukraine have now implemented a number of measures to improve the safety characteristics of the RBMK reactors, but the measures also produce some increases in unit generating costs.

1. The control rod positional set points have also been reset so that all the control rods "dip" into the core at least 1.2 m, with the physical capability to prevent their being withdrawn outside that limit. At the same time the positive scram effect has been eliminated by lowering the rods 0.7 m–1.2 m.
2. The minimum number of control rods in the reactor at any one time has been doubled to 70–80. This limits the influence of the positive void coefficient and ensures a less rapid reactivity insertion.
3. As a longer-term measure the void coefficient has been significantly reduced so that the reactor cannot become prompt-critical. This has been done by increasing the number of fixed absorbers in the core. To compensate for the associated loss of activity, the fuel enrichment has been increased from 2% to

2.4% U-235.

4. Additional instrumentation has been provided to measure subcooling at the inlet to the main circulating pumps.

5. An additional independent "fast" shutdown system with an insertion time of 1–2 seconds has been introduced. The reactor will be automatically tripped without operator intervention if the reactivity margin for control reduces below a preset level.

In addition, wide-ranging improvements in technical management and operator training have been implemented at Chernobyl and the other RBMK reactors.

Given that no accident of such magnitude had previously happened to any nuclear power plant in the world, the coordination and response of the many Soviet recovery services appear to have been exemplary. However, the resource and monetary cost to the Soviet economy is impossible to estimate—it must be at least one order of magnitude greater than the $1 billion for TMI-2.

5.3 HEAVY WATER–MODERATED REACTORS

5.3.1 The NRX Incident

The NRX reactor in Chalk River, Canada, is an experimental reactor, in some respects a forerunner of the present CANDU reactors. It was designed to operate at a full power of 40 MW(t), and the layout of the fuel channels is illustrated in Figure 5.21. Single fuel rods are cooled by light water flowing in an annulus between the rod and a pressure tube, which in turn passes through a calandria tube mounted in a tank of heavy water, which acts as the moderator.

On December 12, 1952. the reactor was undergoing tests at low power. The circulation flow of the light-water coolant was reduced in many of the rods since not much heat was being generated in the fuel. Noting that several red lights indicating withdrawn control rod positions suddenly came on, the supervisor went to the basement and found that an operator was opening valves that caused the control rod banks to rise to their fully withdrawn positions. He immediately closed all of the incorrectly opened valves, after which the rods should have dropped back in. Some of them did, but for unexplained reasons, others dropped in only enough to cause the red lights to turn off. The latter rods were almost completely withdrawn.

From the basement, the supervisor phoned his assistant in the control room, intending to tell him to start the test over again and to insert all the control rods

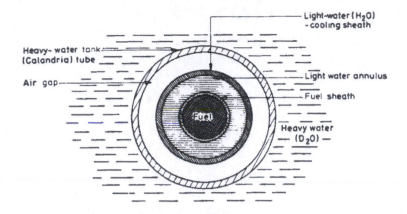

Figure 5.21: Cross section of NRX fuel tube.

by pushing certain buttons. A misunderstanding resulted in the wrong button's being pressed, but the operator in the control room soon realized that the reactor power was rising rapidly and he pressed the "scram" button to trip the reactor. The control rods should then have dropped in under the action of gravity, but many of them did not, and the power continued to climb. After a hurried consultation, it was decided to dump the heavy-water moderator from the calandria tank; this shut down the reactor, but not very quickly since it took some time to drain. The reactor power had peaked between 60 and 90 MW(t).

The increase in power, coupled with the low flow in some of the fuel channels, caused boiling of the light water, which increased the internal pressure and caused the coolant pipes to rupture. The situation was exacerbated by the fact that loss of the light water from the fuel channels gave an increase in reactivity and increased the initial power pulse. Some fuel melting was experienced, and the heavy-water calandria tank was punctured in several places. About 1 million gallons of water containing about 10,000 curies of radioactive fission products had been dumped into the basement of the building.

The core and the calandria, which were damaged beyond repair, were removed and buried, and the site was decontaminated. An improved calandria and core were installed about 14 months after the incident.

The main lesson learned from this incident was that absolute security of control rod operations is mandatory, and modern systems go to great lengths to achieve this. The NRX incident was made worse by the fact that this kind of system has a positive void coefficient, so that the natural event (i.e., the boiling of the water due to heat being input into it) leads to an increase in neutron population.

5.3.2   The Core–Damage Incident at Lucens

The experimental 30-MW(t), carbon dioxide–cooled, heavy water–moderated nuclear power station at Lucens, Switzerland, combined the fuel and coolant of the British Magnox reactors with a heavy-water moderator. The fuel element consisted of a graphite column with seven parallel longitudinal channels (Figure 5.22a). Each channel contained fuel rods made from slightly enriched uranium metal clad in a finned magnesium alloy (Magnox) can (Figure 5.22b). Each fuel element was placed in a Zircaloy pressure tube, closed at the bottom end so that the flow of high-pressure (60 bars) carbon dioxide was directed down the annulus between the graphite column and the pressure tube before passing upward to cool individual fuel rods. The heavy-water moderator was contained in an aluminum alloy calandria tank 3 m in diameter and 3 m high, through which the vertical pressure tubes passed (Figure 5.22c).

On January 21, 1969, an accident occurred that resulted in the destruction of one of the fuel elements and the rupturing of its pressure tube. The carbon dioxide expanded into the moderator tank and, after fracturing its rupture disks, entered the reactor containment, which in this case was an underground cavern, carrying with it fission products and a large fraction of the heavy-water moderator. The reactor was subsequently dismantled.

Postmortem.   The investigations into the causes of the accident were complex and lasted about 10 years (Fritzsche, 1981). The initiating cause of the accident was ingress of water into some of the fuel channels around the edge of the core. This was caused by water leaking from the shaft seals of the carbon dioxide gas circulators. Because the pressure tube was closed at the bottom end, a standing water level was formed in these edge fuel channels when the reactor was shut down. Corrosion at the water-air interface resulted in complete removal of the finning over a short length of the fuel rod.

When the reactor was started up on January 21, 1969, water and corrosion products were blown out of the fuel channel. However, due to the lack of any extended surface in the region of the corrosion damage, the magnesium alloy cladding started to melt (at 640°C). The molten cladding soon ran down the channel and solidified, causing a blockage that prevented coolant flow to that channel. The uranium metal soon reached its melting point (1130°C). The uranium and the magnesium alloy ignited in the carbon dioxide and the molten metals slumped down inside the graphite column. This column, however, was heated nonuniformly.

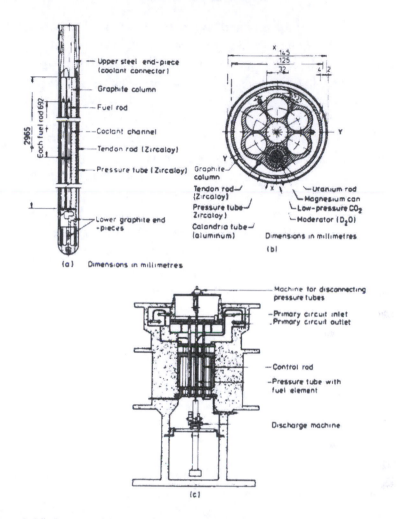

Figure 5.22: Layout of Lucens reactor and fuel element.

The column bowed and contacted the pressure tube, which in turn over-heated and burst open under the action of the coolant gas pressure. Only seconds earlier the reactor had been tripped because of the release of fission products into the coolant gas stream.

Immediately following the pressure tube rupture, the pressure in the moderator tank rose rapidly. At a pressure of 8 bars the bursting disks blew, 0.1 s after the pressure tube rupture, and the expanding CO_2 bubble forced about 1 ton of heavy water out of the moderator tank.

When the pressure tube ruptured, the graphite column also burst apart and the superheated liquid uranium and magnesium metals contacted the pressure tube wall. The Zircaloy wall melted locally and the liquid metal was ejected into the moderator. About 2 kg of the finely dispersed material reacted explosively with the heavy-water moderator. The resulting jet of fire damaged an adjacent pressure tube, which, however, was quenched by returning heavy water before it ruptured. The pressure spike as a result of the chemical explosion reached 16–25 bars and expelled more D_2O from the tank.

Perhaps the most significant aspect of this event was the fact that the ingress of water to the core was not identified. The susceptibility of Magnox cladding to corrosion by water is well known, but the very localized and extensive nature of the corrosion process in removing the finning from the fuel was crucial. The disadvantage of closed-end fuel channels and the separate parallel channels is also to be noted. It was later determined that even if one of the seven flow channels in the graphite column was completely blocked, the flow to that fuel assembly decreased by only 2%.

5.4 GAS-COOLED REACTORS

5.4.1 The Windscale Fire

This accident occurred in one of the large air-cooled reactors (then called "piles") designed for plutonium production and situated at the U.K. Atomic Energy Authority (UKAEA) Windscale works on the northwest coast of England. On October 7, 1957, the reactor was shut down for a routine maintenance operation, which was aimed at releasing the stored ("Wigner") energy deposited in the graphite by atomic displacement, as described in Section 3.3. The accepted practice was to use nuclear heating to bring the graphite moderator up to a temperature where the atoms moved naturally back into their original positions. This process releases further energy, which appears as heat. The heat release is then sufficient to continue the annealing process, and the nuclear heating is discontinued. However, the structure of the Windscale pile was such that pockets of nonannealed graphite presented problems and required a second nuclear heating. At 11 A.M. on October 10, the operators were alerted to the fact that there was a problem by radioactivity monitors, which showed that the activity had increased by a factor of 10 over the normal background level. At 4:30 P.M., visual inspection of the fuel channels revealed that many fuel cartridges were glowing red hot. Attempts to discharge the very hot cartridges failed since they had swelled and jammed in the fuel channels. Further attempts to cool the pile with carbon dioxide during the night of October 10–11 also failed. At 8:55 A.M. on October 11, water was used to cool the very hot fuel, and the core was finally brought to a cold state by 3:20 P.M. on October 12.

Since the reactor was cooled by air, any material released from the burst fuel cartridges was carried in the air stream up to a discharge stack. The stack had a filter system, but it removed only 50% of the particulate emission. It was not effective in removing the noble gases (xenon and krypton) or the volatile iodine-131, and about 20.000 curies of iodine were released to the atmosphere.

Investigations after the accident suggested that the second nuclear heating was applied too rapidly and, as a result, one of the fuel cartridges burst. Oxidation of the uranium in this burst fuel cartridge caused a fire, including combustion of the surrounding graphite moderator. The burning of the graphite released further energy in the zone of the core around the original point of ignition, and by the evening of October 10, 150 channels containing approximately 8 tons (8000 kg) of uranium fuel were on fire. Showing very considerable courage, the operators created a firebreak by discharging the fuel cartridges from the channels adjacent to the combustion zone. When water was finally used to cool the channels, there was a recognized considerable risk of explosion and thus a greater release. The station was placed on emergency during this procedure.

This early form of reactor is obviously very different from a modern power station. The use of metal fuel led to the combustion, which initiated a graphite fire, which was kept going by the continuing flow of air through the reactor. However, the incident is of particular interest in nuclear safety analysis because of the iodine release, which was much greater than that which occurred, for example, at Three Mile Island.

The filters placed in the stack, which held back 50% of the radioactive iodine, the released strontium, and the released cesium, were an afterthought and the result of the insistence of Dr. (later Sir) John Cockcroft of the UKAEA. They were known colloquially as "Cockcroft's follies." Although these filters were clearly very helpful in limiting the release, their design was inadequate to trap the volatile fission products.

After the accident, milk supplies were monitored; radioactive iodine can easily find its way into milk by deposition on grassland and ingestion by cows. The sale of milk derived from herds in that part of England surrounding the Windscale plant was stopped for about 6 weeks.

The consequences of the Windscale incident have been studied by the National Radiological Protection Board. It has been estimated that ~30 additional cancer deaths may have occurred in the general public, representing a 0.0015% increase in the cancer death rate (in other words, over the period when these 30 deaths may have occurred, 1 million deaths from cancer would have occurred in the exposed population).

5.4.2   The Fuel Meltdown at St. Laurent

The St. Laurent plant of Electricité de France is a 500-MW(t) Magnox reactor that was first brought into operation in January 1969. The reactor is fueled on load and the machine that carries this out is called a *charging machine*. We shall consider the use of these machines in Chapter 7 in discussing the handling of fuel elements subsequent to their period in the reactor core. The charging machine is a very large device that is computer-controlled to move about the top of the reactor and position itself properly over each access port to unload and load the fuel. Figure 5.23 illustrates the layout of the St. Laurent reactor.

During the midnight shift on October 17, 1969, with the reactor near full power, a normal loading and unloading operation was in progress. Graphite plugs that had been placed temporarily in one of the fuel channels in the core were being replaced by fuel. The charging machine had unloaded the graphite plugs from the core into its empty storage chambers and had loaded fuel into the core from two of its full chambers, but then it stopped. Three full chambers of fuel elements are required to load one fuel channel in the core completely, and each chamber contains four elements. When the charging machine stopped, the operator overrode the automatic system, and after a series of manual operations, he accidentally charged a flow restriction device into the channel instead of a fuel element. These flow restrictors were used to control the gas flow to individual channels. The loading of a flow restrictor into this particular channel so reduced the flow that the fuel elements were inadequately cooled.

Some of the fuel elements in the affected channel heated up beyond their melting point, and the molten fuel flowed out of the channel onto the diagrid below (Figure 5.23). This released radioactive fission products, set off alarms, and activated a reactor trip. The molten fuel (about 50 kg) was still still contained within the massive concrete structure; hence, little, if any, radioactivity was released outside the structure and there were no injuries. However, a year was needed to complete the cleanup operations and restart the reactor. Modifi-

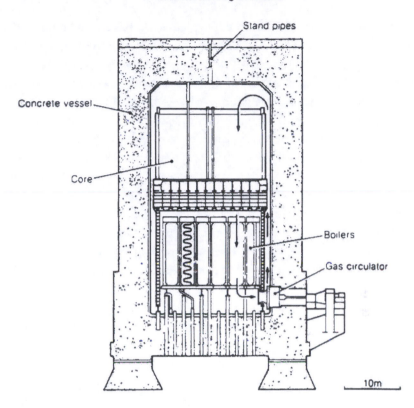

Figure 5.23: Reactor at St. laurent.

cations to the machine were made, and it is no longer so easy to override the automatic system and bring the machine into manual control.

This accident and a similar one at the British Chapelcross reactor in Scotland again demonstrate the importance of carefully matching the heat removal and heat input characteristics for the system as a whole and for each component part. Again, the scope for operator error is noted, and this has necessitated steps to reduce the scope.

5.4.3   Seawater Ingress in the Hunterston B AGR Station

This incident occurred soon after the initial commissioning of the advanced gas-cooled reactors at Hunterston in Scotland. On October 2, 1977, the B2 reactor was shut down for modifications to the plant. On October 11, the carbon dioxide gas pressure was being reduced when alarms, instruments reading, and gas samples began to show excessive moisture in the reactor coolant gas. Subsequently, it was discovered that about 8000 liters of seawater had entered the reactor vessel. Damage to the insulation in the annulus below the boilers was extensive. It had to be completely replaced and the reactor was out of service for about 28 months. The repair work cost £13 million (Gray et al., 1981).

At first it seems incredible that a large amount of seawater could enter the pressure vessel of a gas-cooled reactor. The circumstances were these. Figure 5.24 shows the gas circulator cooling system. During initial commissioning of the reactor in April 1977, the demineralized water in the cooling circuit for the seals on one of the circulators was found to be acidic due to the presence of carbon dioxide. Carbon dioxide was entering the cooling water through a crack in a seal weld. In order to allow the reactor to run until its planned shutdown in October, it was decided to continue the commissioning phase of the operation and run the acidic water to waste via a temporary connection to the reactor seawater cooling system, thereby avoiding corrosion of the circulator cooling system.

When the gas pressure was reduced below the seawater cooling system pressure, a flow path for the seawater was established. This would not have happened if the isolating valves in the temporary drain connection, which had earlier been logged as shut, had in fact been shut. Actually, they were partly open.

This incident points to the dangers of temporary modifications made without

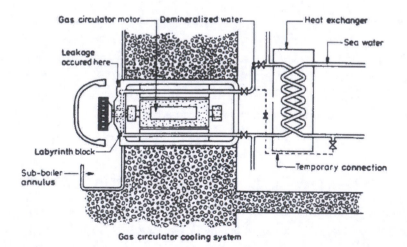

Figure 5.24: Hunterston B gas circulator cooling system.

full analysis of all the implications and to the importance of positive indication of valve positions.

5.4.4 Fuel Damage during Charging at the Hinkley Point B AGR

The advanced gas-cooled reactors are designed to be refueled while in operation. Initial on-load refueling operations with the first two AGRs at Hunterston and Hinkley Point were confined to the charging of fuel into channels in which dummy fuel assemblies had been loaded when the reactor was first charged with fuel. By November 1978, some 15 fuel assemblies at Hinkley and 20 at Hunterston had been charged on-load into these so-called vacancy channels.

On November 19, 1978, a fuel assembly was being withdrawn from channel 4K05 on Hinkley Point B reactor R4. The assembly was raised about 10 ft and then snagged, and the charge machine hoist tripped out on overload. Subsequently, it was successfully raised into the charge machine. Visual examination of the connected string of fuel elements withdrawn from this reactor channel (the *stringer*) showed the graphite sleeves surrounding the third, fourth, and fifth elements to be severely damaged. Damage to the graphite sleeve resulted in the fuel elements above the damaged sleeve being starved of coolant and thus overheating, resulting in failure of some of the fuel "pins" that made up the element. Subsequently, a large portion of graphite sleeve from element 4 was recovered from the reactor during a statutory in-reactor inspection. The level of radiation from the sleeve suggested that it was never in the reactor core and that the damage occurred during the loading process. The damaged assembly had been loaded into a vacancy channel at 82% power earlier in the year. The incident caused doubts about the safety of refueling AGRs at power, and an embargo was placed on on-load refueling. A program of investigations was begun to establish the cause of the problem.

When the fuel is being lowered into the reactor, it receives considerable buffeting from the very high gas flow through the empty channel. It is believed that small cracks may have been present in a number of fuel element sleeves and that the sleeve of element 4 cracked further due to the pressure differential across the sleeve during on-load refueling. Techniques have been developed to detect cracks in sleeves, and these and other improvements have been incorporated into the AGRs. On-load refueling has been resumed at low power.

5.5 LIQUID METAL–COOLED FAST REACTORS

5.5.1 The EBR–1 Meltdown Accident

The U.S. Experimental Breeder Reactor I (EBR-1) had the distinction of being the first reactor to generate electricity. Construction of the reactor began in 1948, and electric power production started in December 1951. The reactor was designed for a thermal output of 1 MW(t) and an electrical power output of 200 kW(e). Of course, the power production was more for demonstration than for economic viability.

The core of the reactor is illustrated schematically in Figure 5.25a. During its lifetime, the reactor was operated with four different core configurations, all with fuel in metallic form. The first three cores were of highly enriched uranium, consisting mainly of U-235. The second core had a uranium-zirconium alloy fuel containing 2% zirconium. The fuel pins were 1.25 cm in diameter, and 217 pins in a triangular array were mounted in a central hexagon 19 cm across, forming the core of the reactor. The small size of this core illustrates the great compactness of liquid metal–cooled fast reactors. Around the central U-235 region there was a blanket region containing natural uranium rods, as shown in Figure 5.25a. The coolant for the reactor was a sodium-potassium mixture (NaK) that is liquid at room temperature (see Chapter 3).

With the second core, power oscillations were observed at very low core flows. In an experiment to examine this effect beginning on November 29, 1955, with the core flow totally stopped and certain safety interlocks cut out, power was rapidly raised in order to determine the magnitude of a previously observed increase in reactivity with temperature. It had been intended to terminate the experiment with the fuel temperature at 500°C, but through the combination of this temperature effect and an operator error, the temperature rose to more than 720°C. At this temperature the uranium metal fuel and the stainless steel can begin to interact, leading to the melting of about 40% of the core, but without explosion, plant damage, or radiation hazard.

As explained in Chapter 4, bringing the pins closer together in a fast reactor causes an increase in reactivity or neutron population. The mechanism by which the EBR-1 core meltdown occurred was related to this. It was possible for the rods to bow as illustrated in Figure 5.25b, and this gave an increase in reactivity that was self-propagating as the increased temperatures increased the amount of bowing. This accounted for the temperature effect that was being investigated at the time and that was subsequently explained theoretically. The core of EBR-1 was later removed and replaced by another core designed to eliminate the bow-

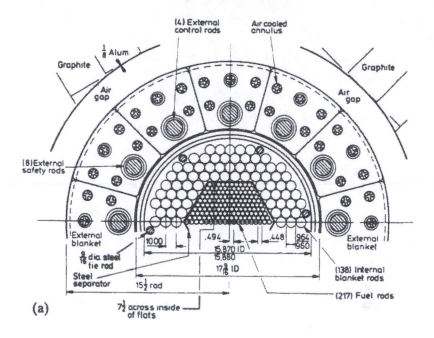

(a)

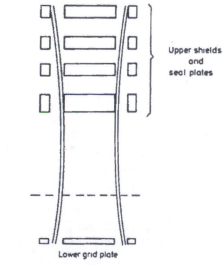

(b)

Figure 5.25: The EBR-1 meltdown incident.

ing effect by the use of spacer ribs. The expansion of the ribbing with increasing temperature causes the core to expand, giving a negative rather than the previously observed positive temperature coefficient of reactivity.

The EBR-1 reactor, which was finally shut down in December 1963, gave information of great value related to the design of fast reactors. Now all fast reactor cores are designed with significant amounts of restraint so that they always have a negative temperature coefficient of reactivity. In fact, it may be possible in the future to design fast reactor cores that are inherently safe in that they expand to switch off the nuclear reaction even if the control rods fail to actuate. This is one of the features of fast reactors that make them in some respects even safer than thermal reactors.

5.5.2   Fuel Melting Incident at the Enrico Fermi 1 Fast Breeder Reactor

The Enrico Fermi reactor was a sodium-cooled fast breeder demonstration reactor, producing 200 MW(t) [61 MW(e)]. The plant was located near Lagoona Beach, Michigan, and started operation in 1963. After extended low-power operation, power raising took place during 1966. When this was being done, it was noted that the coolant temperatures above 2 of the 155 fuel assemblies (clusters of fuel rods) were higher than normal and the temperatures above another assembly were lower than normal.

The reactor was shut down, and the fuel assemblies were rearranged in the core to determine whether these abnormal temperatures were dependent on location in the core or were characteristic of the fuel assemblies themselves.

On October 5, 1966, the rise to the selected power level [67 MW(t)] for these tests on the rearranged fuel elements was begun. At about 3 P.M., with the reactor at a power level of 20 MW(t), the reactor operator observed a control signal, indicating that the rate of change of neutron population was erratic. The problem had been experienced before and was thought to be due to random electrical fluctuations in the control system. The reactor was placed on manual control, and when the instability disappeared, automatic control was again selected and the increase in power resumed.

At 3:05 P.M., with the reactor power at 27 MW(t), the erratic signal was again observed. Shortly after that it was noted that the control rods were withdrawn farther than normal. A check of the core exit temperatures showed that the outlet temperatures from two subassemblies were abnormally high at 380 and 370°C (715 and 695°F), compared with a mean bulk outlet temperature of 315°C (600°F).

At 3:09 P.M., alarms occurred from the ventilation monitors in the upper building ventilation exhaust ducts. The building was automatically isolated—no one was inside at the time—and a radiation emergency was announced. The reactor power increase was stopped at 31 MW(t), and a power reduction was started. By 3:20 P.M., the power had decreased to 26 MW(t) and the reactor was manually tripped and shut down indefinitely. Over the next year, many of the assemblies were removed and examined, and it was found that the bulk of the fuel in two of the fuel assemblies had melted. It was not until the end of the examination period that the cause of the accident was discovered. The cause was relatively trivial. Below the core, six small Zircaloy plates had been installed to guide the flow of sodium into the upward direction. One of these Zircaloy plates had broken loose and blanked off the entry to a few subassemblies, causing almost total flow starvation.

The damage to the reactor was repaired with a specially designed remote handling tool, and the reactor reached full power output again in October 1970, 4 years after the accident.

Although the Enrico Fermi accident led to no injury or release of activity outside the containment shell, 10,000 curies of fission products were released to the circulating sodium coolant. The accident focused attention on the potential problems of flow blockages caused by foreign bodies within the circulating sodium. In particular, any part of the reactor that may be susceptible to vibration damage, causing the release of foreign material, must be carefully evaluated. In the design of modern reactors, very thorough flow testing of the various components is carried out. It is noteworthy that the zirconium plates were added at a very late stage in the design and may not have had the same level of quality assurance as the other components in the Enrico Fermi reactor. Late "fix-ups" of this kind and of the kind that occurred at Hunterston must be avoided.

The damage to the fuel assemblies did not propagate to adjacent fuel assemblies, and the evidence from this incident that the accident did not escalate was encouraging.

5.6 THE INTERNATIONAL NUCLEAR EVENT SCALE (INES)

One lesson stemming from the Chernobyl accident was the need for prompt dissemination to the public of the safety significance of an event at a nuclear installation. A similar need in other areas is filled by an appropriate *scale*, for example, the Richter scale for earthquakes and the Beaufort scale for winds.

In 1990 the International Atomic Energy Agency (IAEA) introduced a seven-level scale designed to allow prompt classification of such events. The levels, their descriptions, and detailed criteria are shown in Figure 5.26. Three criteria are applied:

Levels 3–7 relate to the extent of releases of radioactivity off-site.
Levels 2–5 relate to the extent of on-site contamination or exposure.
Levels 1–3 relate to the extent to which the defense-in-depth philosophy has been challenged.

Each of the incidents described in this chapter has been evaluated using the INES scale to provide a best estimate of the incident. The resulting classification is given in Table 5.2.

The International Nuclear Event Scale
For prompt communication of safety significance

Figure 5.26: Diagrammatic representation of the International Atomic Energy Agency scale for events at nuclear installations.

Table 5.2 • The International Nuclear Event Scale (for prompt communication of safety significance)

Level	Descriptor	Criteria	Examples
Accidents			
7	Major accident	• External release of a large fraction of the reactor core inventory typically involving a mixture of short- and long-lived radioactive fission products (in quantities radiologically equivalent to more than tens of thousands terabecquerels of iodine-131).	Chernobyl, USSR 1986
		• Possibility of acute health effects. Delayed health effects over a wide area, possibly involving more than one country. Long-term environmental consequences.	
6	Serious accident	• External release of fission products (in quantities radiologically equivalent to the order of thousands to tens of thousands of terabecquerels of iodine-131). Full implementation of local emergency plans most likely needed to limit serious health effects.	
5	Accident with off-site risks	• External release of fission products (in quantities radiologically equivalent to the order of hundreds to thousands of terabecquerels of iodine-131). Partial implementation of emergency plans (e.g., local sheltering and/or evacuation) required in some cases to lessen the likelihood of health effects.	Windscale, UK 1957
		• Severe damage to large fraction of the core due to mechanical effects and/or melting.	Three Mile Island, USA 1979
4	Accident mainly in installation	• External release of radioactivity resulting in a dose to the most exposed individual off-site of the order of a few millisieverts.[a] Need for off-site protective actions generally unlikely except possibly for local food control.	
		• Some damage to reactor core due to mechanical effects and/or melting.	Saint Laurent, France 1980
		• Worker doses that can lead to acute health effects (of the order of 1 Sievert).[b]	
Incidents			
3	Serious incident	• External release of radioactivity above authorized limits, resulting in a dose to the most exposed individual off-site of the order of tenths of a millisievert.[a] Off-site protective measures not needed.	
		• High radiation levels and/or contamination on-site due to equipment failures or operational incidents. Overexposure of workers (individual doses exceeding 50 millisieverts).[b]	

Table 5.2 continued

Level	Descriptor	Criteria	Examples
		• Incidents in which a further failure of safety systems could lead to accident conditions, or a situation in which safety systems would be unable to prevent an accident if certain initiators were to occur.	Vandellos, Spain 1989
2	Incident	• Technical incidents or anomalies which, although not directly or immediately affecting plant safety, are liable to lead to subsequent reevaluation of safety provisions.	
1	Anomaly	• Functional or operational anomalies which do not pose a risk but which indicate a lack of safety provisions. This may be due to equipment failure, human error, or procedural inadequacies. (Such anomalies should be distinguished from situations where operational limits and conditions are not exceeded and which are properly managed in accordance with adequate procedures. These are typically "below scale.")	
Below scale / zero	No safety significance		

Source: International Atomic Energy Agency, April 1990.

[a] The doses are expressed in terms of effective dose equivalent (whole body dose). Those criteria, where appropriate, also can be expressed in terms of corresponding annual effluent discharge limits authorized by National authorities.

[b] These doses also are expressed, for simplicity, in terms of effective dose equivalents (Sieverts), although the doses in the range involving acute health effects should be expressed in terms of absorbed dose (Grays).

Table 5.3 shows the ratings of the various incidents discussed in this chapter in terms of the INES scale. This table also shows how each of the safety principles (the Three Cs—see Section 5.1) were met in each case and whether defense in depth was effective.

Table 5.3 • Nuclear Reactor Incidents

| | SAFETY PRINCIPLES (THREE CS) | | | | International Nuclear Event Scale Rating |
	Control the Reaction	Cool the Fuel	Contain the Radioactivity	Defense in Depth	
Light water-cooled reactors					
SL1	x	—	✓	✓	4
Millstone 1	✓	✓	[?]	✓	3
Browns Ferry 1 and 2	✓	[✓]	[✓]	✓	3
Three Mile Island-2	✓	x	✓	✓	5
Ginna	✓	✓	[?]	✓	2
Mihama-2	✓	✓	[?]	✓	2
Chernobyl	x	—	x	x	7
Heavy water-cooled reactors					
NRX	x	—	[?]	✓	4
Lucens	✓	x	✓	✓	4
Gas-cooled reactors					
Windscale	✓	x	x	x	5
St. Laurent	✓	x	✓	✓	4
Hunterston B	✓	✓	[?]	✓	1
Hinkley Point B	✓	[?]	✓	✓	2
Liquid metal-cooled reactors					
EBR-1	x	x	✓	✓	4
Enrico Fermi	✓	x	✓	✓	4

x Safety principle violated
✓ Safety principle complied with

REFERENCES

ApSimon, H.M., et al. (1986). "An Initial Assessment of the Chernobyl-4 Reactor Release Source." *J. Soc. Radiol. Prot.* 6 (3).

Arnold, L. (1992). *Windscale 1957: Anatomy of a Nuclear Accident.* Macmillan, London, 235 pp.

Bertini, H.W., et al. (1980). "Descriptions of Selected Accidents That Have Occurred at Nuclear Reactor Facilities." *ORNL/NSIC-176*, April 1980, Oak Ridge National Laboratory, Oak Ridge, Tenn.

Fritzsche, A.F. (1981). "Accident at the Experimental Nuclear Power Station in Lucens." *Nuc. Safety* 22 (1): 87–100.

Gittus, J.H., et al. (1988). *The Chernobyl Accident and Its Consequences,* 2d ed. Report NOR 4200, U.K. Atomic Energy Authority.

Gray, J.L., et al. (1981). "The Repair of an Advanced Gas Cooled Reactor at Hunterston 'B' Power Station following an Ingress of Sea Water." *Proc. Inst. Mech. Eng.* 195 (9): 87–99.

Holloway, N.J. (1993). "The Safety of RBMK Nuclear Power Plants." *Nuclear Engineer* 34 (September–October 1993):135–41.

INSAG (International Nuclear Safety Advisory Group) (1986). *IAEA Vienna.* Summary Report on the Post Accident Review Meeting on the Chernobyl Accident, INSAG-1.

——— (1992). *The Chernobyl Accident: Updating of INSAG-1, IAEA Vienna.*

Livens, Dr. F., Institute of Terrestrial Ecology (Merlewood Research Station). Private communication.

Mosey, D. (1990). *Reactor Accidents: Nuclear Safety and the Role of Institutional Failure.* Nuclear Engineering International Special Publications, 108 pp.

Read, P.P. (1993). *Ablaze: The Story of Chernobyl.* Secker and Warburg, London, 478 pp

EXAMPLES AND PROBLEMS

1 Decay heat removal using PORVs

Example: Following the TMI accident, a utility was considering the possibility of increasing the number of PORVs in its 4000-MW(t) PWR system to allow release (in the form of steam) of the full decay energy at 100 s from shutdown. Assuming a flow area for each valve of 0.002 m², how many valves would be required?

Solution: After 100 s, the decay heat rate may be estimated from Table 2.2 and is

$$3.2 \times 4000 / 100 = 128 \, \text{MW}$$

The flow area required can be estimated assuming a release rate of 17,000 MW/m² (see Section 4.3.2).

Thus
$$\text{Flow area} = \frac{128}{17,000} = 0.0075 \, \text{m}^2$$

and four PORVs would be required.

Problems: A 3000-MW(t) PWR has two PORVs, each with a flow area of 0.0015 m². Would these valves be sufficient to allow release of decay energy from the reactor vessel in the form of steam, and consequent maintenance of fuel cooling by "feed-and-bleed" operation, at 1 h from shutdown?

2 Evaporation of coolant

Example: Following a small-break loss-of-coolant accident, the fuel of a 3800-MW(t) PWR has become uncovered and the top half of the fuel is dry. What is the rate at which the core is becoming uncovered at 1 h after shutdown, assuming a mean void fraction in the wetted region of 0.5? Also assume that the fuel occupies 40% of the core volume, that the core diameter is 3.6 m, and the core length 4 m, and that the heat flux is uniform in the core. The system pressure during the uncovery period was 85 bars.

Solution: The volume of water per meter length of the core in the wetted region is given by

$$\text{Cross-sectional area of core} \times (1 - \text{void fraction})$$
$$\times (1 - \text{fractional area occupied by fuel})$$
$$= \left(\frac{\pi}{4} \times 3.6 \times 3.6\right) \times (1 - 0.5) \times (1 - 0.4) = 3.054\,\text{m}^3/\text{m}$$

The heat release rate to water from the submerged half of the fuel at 1 h from shutdown is given (using Table 2.2) by

$$\frac{3800 \times 10^6}{2.0} \times \frac{1.4}{100} = 2.66 \times 10^7\,\text{W}$$

Evaporation rate of water

$$= \frac{\text{heat release rate}}{\text{latent heat of evaporation of water at 85 bars}}$$
$$= \frac{266 \times 10^7\,\text{W}}{1.40 \times 10^6\,\text{J/kg}} = 19\,\text{kg/s}$$

Volume evaporation rate $\quad = \dfrac{19\,\text{kg/s}}{\text{density of water}} = \dfrac{19\,\text{kg/s}}{713\,\text{kg/m}^3}$

$$= 0.0266\,\text{m}^3/\text{s}$$

Uncovery rate $\quad = \dfrac{\text{volume evaporation rate}}{\text{volume of water per unit length}}$

$$= \frac{0.02665\,\text{m}^3/\text{s}}{3.054\,\text{m}^3/\text{s}} = 0.0873\,\text{m/s}$$
$$= 31.4\,\text{m/h}$$

It is now necessary to iterate to ensure consistency with only half of the core being uncovered in 1 h.

Problem: For the reactor core described in the example, the heat flux would not in practice be uniformly distributed. Rather the flux profile along the core length follows a law that would typically be of the following form:

$$\dot{q} = F\dot{q}_{av} \sin\frac{\pi(z+a)}{L+2a}$$

where $\dot{q}$ is the local heat flux, $\dot{q}_{av}$ the average heat flux, z is the distance from the bottom of the core, L is the core length, and a is a constant. F is a form factor (ratio of peak to average heat flux). Assuming $F = 1.4$ and $a = 0.3$, calculate the total time required to totally uncover the core described in the example. Assume a constant heat input equivalent to that occurring 1 h after shutdown, that the core is initially just filled with a steam mixture water with 50% void fraction, and that the void fraction remains constant during the uncovery. Also, plot the movement with time of the mixture level.

3 Fuel blockage in a fast reactor

Example: Calculate the location and magnitude of the peak clad temperature in the peak rated channel of a fast reactor under normal flow conditions. Would a blockage leading to a 50% reduction in flow lead to the fuel elements exceeding the creep limit of 670°C, above which ballooning of the cans would occur? In the calculations, assume a 3300-MW(t) reactor having hexagonal fuel assemblies, which each have 325 fuel pins 5.84 mm in diameter with the distance across the faces of the hexagon being 135 mm. The normal mass rate of flow through each subassembly (M) is 39kg/s, and the core length is 1 m. Liquid sodium enters the core region at 370°C. In the core region the peak fuel rating in the highest-rated fuel assembly is 44 kW/m* and (for the purposes of this present calculation**), assume that the local rating r is given by

$$r = r_{max} \sin\frac{\pi z}{L} = 44\sin\frac{\pi z}{L}$$

where z is the distance from the beginning of the core and L is the core length. Assume a heat transfer coefficient x between the fuel and the sodium of 55,000 W/m² K at the full flow conditions and 32,000 W/m² K at 50% flow. Assume that the sodium has a specific heat capacity (c_p) of 1275 J/kg K.

Solution: The total heat generation rate (Q_L) in this assembly is:

$$\dot{Q}_L = 325\times\int_0^L r_{max}\sin\frac{\pi z}{L}dz$$

$$= 325\times\left[-r_{max}\frac{L}{p}\cos\frac{\pi z}{L}\right]_0^L r$$

$$= 325\times\frac{2Lr_{max}}{\pi} = 9.10\times10^6 \text{ W}$$

---

* The difference between this value and the value of 27 k W/m given in Table 2.3 is that the figure in the table was an *average* rating including those parts of the fuel in the blanket and outer core regions.

**Note: The equation for flux profile implies that the flux goes to zero at the bottom and top of the core. This simplifies the calculation, but the actual profile would go to a finite rating at the extremities of the core, and, indeed, there is a finite generation of heat in the blanket region above and below the core.

The temperature rise ΔT_L across the element under full flow conditions is thus

$$\Delta T_L = \frac{\dot{Q}}{Mc_p} = \frac{9.10 \times 10^6}{39 \times 1275} = 183 \text{ K}$$

The temperature difference ΔT_w between the fuel element and the sodium is given by

$$\Delta T_w = \frac{\text{heat flux from the fuel element surface}}{\text{heat transfer coefficient}}$$

$$= \frac{\dot{q}}{\alpha} = \frac{r}{\pi D \alpha} = \frac{r_{max}}{\pi D \alpha} \sin \frac{\pi z}{L}$$

where D is the fuel pin diameter.

The fuel pin surface temperature is thus

$$T_w = T_s + \Delta T_w$$

where T_s is the sodium temperature that is given at distance z by the heat balance.

$$T_s = 370 + \frac{\dot{Q}_z}{\dot{M}c_p}$$

where Q_z is the total heat generated in the fuel assembly from the inlet to position z. Thus

$$T_s = 370 + \frac{325}{\dot{M}c_p} \int_0^L r_{max} \sin \frac{\pi z}{L} d_z$$

$$= 370 + \frac{325}{\dot{M}c_p} \left(-r_{max} \frac{L}{\pi} \cos \frac{\pi z}{L} \right)_0^z$$

$$= 370 + \frac{325}{\dot{M}c_p} \frac{r_{max}L}{\pi} \frac{r_{max}L}{\pi} \cos \frac{\pi z}{L}$$

Thus, the pin surface temperature varies with distance along the element as follows:

$$T_w = T_s + \Delta T_w$$

$$= 370 + \frac{325}{\dot{M}c_p} \left(\frac{r_{max}L}{\pi} - \frac{r_{max}L}{\pi} \cos \frac{\pi z}{L} \right)$$

$$+ \frac{r_{max}L}{\pi D \alpha} \sin \frac{\pi z}{L}$$

When T_w is a maximum, $dT_w/dz = 0$. Thus

$$\frac{dT_w}{dz} = \frac{325}{\dot{M}c_p} \frac{r_{max}L\pi}{\pi L} \sin \frac{\pi}{L} + \frac{r_{max}L}{\pi D \alpha} \frac{\pi}{L} \cos \frac{\pi z}{L}$$

and for this condition

$$\frac{325}{\dot{M}c_p}\sin\frac{\pi z}{L} = -\frac{1}{D\alpha}\cos\frac{\pi z}{L}$$

or

$$\tan\frac{\pi z}{L} = \frac{-\dot{M}c_p}{325 D\alpha}$$

$$z = \frac{L}{\pi}\tan^{-1}\left(\frac{-\dot{M}c_p}{325 D\alpha}\right)$$

$$= \frac{1}{\pi}\tan^{-1}\left(\frac{-39\times1275}{325\times5.84\times10^{-3}\times5.5\times10^{4}}\right)$$

$$= \frac{1}{\pi}\tan^{-1}(-0.4763)$$

$$= \frac{2.697}{\pi} = 0.8585\,\text{m}$$

At $z = 0.8585$ m, the maximum pin surface temperature for normal flow conditions is given by

$$T_w = 370 + \frac{325 r_{max} L}{\dot{M}c_p \pi}\left(1 - \cos\frac{\pi z}{L}\right)$$

$$+ \frac{r_{max}L}{\pi D\alpha}\sin\frac{\pi z}{L}$$

$$= 370 + \frac{325\times44,000\times1}{39\times1275\times\pi}(1+0.9028)$$

$$+ \frac{44,000\times1}{\pi\times5.84\times10^{-3}\times5.5\times10^{4}}\times0.4300$$

$$= 370 + 174.2 + 18.7 = 562.9$$

Thus, the peak clad temperature is normally well below the creep limit of 670°C. For a flow reduction of 50%, the peak clad temperature occurs at a distance z from the inlet given by

$$z = \frac{L}{\pi}\tan^{-1}\left(-\frac{\dot{M}c_p}{325 D\alpha}\right)$$

$$= \frac{1}{\pi}\tan^{-1}\left(\frac{-39\times1275}{325\times5.84\times10^{-3}\times3.2\times10^{4}}\right)$$

$$= 0.8763\,\text{m}$$

The clad temperature at this position is given by

$$T_w = 370 + \frac{325 r_{max} L}{\dot{M}c_p \pi}\left(1 - \cos\frac{\pi z}{L}\right)$$

$$+ \frac{r_{max}L}{\pi D\alpha}\sin\frac{\pi z}{L}$$

$$= 370 + \frac{325 \times 44,000 \times 1}{39 \times 1275 \times \pi}(1 + 0.92555)$$

$$+ \frac{44,000 \times 1}{\pi \times 5.84 \times 10^3 \times 3.2 \times 10^4} \times 0.3789$$

$$= 370 + 352.6 + 28.4$$

$$= 751.0°C$$

Thus, a flow blockage leading to a 50% reduction in flow would lead to the peak clad temperature in excess of the creep limit of 670°C and would be unacceptable.

Problem: If, for the fast reactor described in the example above, the flow reduction due to blockage was even greater than 50%, boiling of the sodium would ultimately begin at the fin surfaces. The boiling point of sodium would be required to initiate boiling; calculate the flow reduction that would be required to cause boiling to start on the peak-rated fuel assembly. Also calculate the position on the fuel assembly at which such boiling would be initiated.

BIBLIOGRAPHY

Atomic Energy Office (1957). *Accident at Windscale No. I Pile; 10 October 1957.* HMSO, London, 22 pp.

Cantelon, P.L., and R.C. Williams (1980). *A History.* National Technical Service, The Department of Energy at Three Mile Island, 217 pp.

Construction, Commissioning and Operation of Advanced Gas-Cooled Reactors. Proceedings of a Conference (1977). Institution of Mechanical Engineers, London, 135 pp.

Fast Reactor Safety Technology, Proceedings of a Meeting (1979). Seattle, August 19–23. American Nuclear Society, LaGrange Park, Ill., 5.vols.

Hart, G. (1980). *Nuclear Accident and Recovery at Three Mile Island.* Report prepared by the Subcommittee on Nuclear Regulation for the Committee on Environment and Public Works, 96th Cong. 2d sess. U.S. Government Printing Office, Washington, D.C., 430 pp.

"Hinkley Point B: Special Survey (with Pullout Cut-away Drawing)" (1968). *Nucl. Eng.* 13 (147): 652–68.

Hu, T.W., and K.S. Slaysman (1982). "Health-Related Economic Costs of the Three Mile Island Accident." Presented at the American Public Health Association (APHA) Meeting, Montreal, November 15–17, 28 pp.

Kemeny, J.G. (1979). *Report on the President's Commission...The Need for Change: The Legacy of TMI.* Pergamon, Elmsford, N.Y., 179 pp.

Parker, Hon. Justice (1978). *A Report on the Windscale Enquiry:* vol. 1, Report and Annexes 3–5; vol. 2, List of Appearances, List of Documents; vol. 3, Index to the Report. HMSO, London.

Sills, D.L. (1982). *Accident at Three Mile Island: The Human Dimensions.* Westview, Boulder, Colo., 258 pp.

6

Postulated Severe Accidents

6.1 INTRODUCTION

In Chapters 4 and 5 we discussed the means by which loss-of-coolant accidents (LOCAs) could occur and the ways in which reactors must be designed to cope with these extremely unlikely events. We also discussed in Chapter 5 a number of actual incidents in reactors where a failure of cooling occurred with consequent overheating and fuel damage. Many of these conditions were anticipated in the design, but some actually went beyond the design basis. In all cases, except Chernobyl and Windscale, the "defense in depth" approach to nuclear reactor design was effective in limiting the public consequences of the accident. However, it is important to consider what might be involved in extremely severe accidents having the common characteristics of leading to partial or complete meltdown of the fuel within the reactor.

In classifying operational states in Section 4.1, we considered a series of transient events in reactors ranging from operational transients to limiting fault conditions. Even in a limiting fault condition, the reactor is designed so that there is no loss of coolability of the core over protracted periods. However, one can postulate a situation in which the emergency core cooling system (ECCS) itself fails and no other cooling system is available. Another possibility would be loss-of-site power over a long period, coupled with inability to actuate the alternative power sources (normally on-site diesel engines). A third possibility is that of unpredicted operator faults, which may lead, as at Three Mile Island, to conditions beyond those designed for as limiting.

Also in Section 4.1 we described the concept of containment and the various barriers preventing the release of activity:

- The matrix of the fuel itself and the cladding around the fuel
- The reactor pressure vessel
- The containment building or system

We explained that the whole purpose of the safety system provided on a reactor was to ensure that these separate barriers are *not* challenged and all remain intact. That is embodied in the safety case. But suppose these systems *are* degraded in some way or are inoperative and their purpose is *not* achieved. What then?

To answer this question, it is therefore informative to examine how each barrier might be challenged and the failure mode and consequences that might result. That, in turn, will lead to consideration of design measures to limit the failure or mitigate the consequences.

6.2 POSTULATED SEVERE ACCIDENTS IN WATER-COOLED REACTORS

6.2.1 Core Damage

Essentially, the first barrier, that of the fuel matrix and its cladding, can be challenged in one of two ways: loss of cooling or increase of power. First, loss of effective cooling of the fuel can lead to overheating as happened at TMI-2. Alternatively, a significant increase of neutron population (or reactor power) can result in excess energy deposition within the fuel, leading to fuel expansion and melting and consequent failure of the cladding. This can occur in spite of apparently adequate cooling. The accident at Chernobyl is an extreme example of this class of fuel failure.

Let us concentrate initially on the consequences of a loss-of-cooling situation. There are many ways this could develop with the primary circuit at either high pressure or depressurized, and on a time scale of a few seconds to a few hours.

The progressive failure of the fuel can be summarized as follows:

1. As the fuel canning material increases in temperature, it will either burst or under some circumstances swell because of the gas pressure inside it. This may lead to a restriction of the coolant flow between and around the fuel elements and make more difficult the problem of cooling them. This factor is, in fact, taken account of in the design of fuel for water-cooled reactors, and it has been shown that blockages of up to 90% can be coped with.

2. As the fuel temperatures rise, so the volatile fission products are released and a temperature is reached (1200–1400°C) at which the first signs of molten material in the core begin to be observed. The melting process is very complex with the formation of eutectics and occurs most rapidly in the regions of the core that have had the highest neutron flux (and therefore the highest concentration of fission products whose decay is causing the heating). The

grids that hold the fuel together also melt around 1400°C, followed by the control rods passing through the fuel.

3. At core temperatures above 1100°C the steam reacts with the zirconium can, destroying the can. The reaction is exothermic, that is, the chemical reaction itself releases additional heat. As the temperature increases, so the reaction rate increases and at high temperatures the chemical reaction can contribute as much or even more heat than the fission decay process. The chemical reaction produces hydrogen, which we shall see is a potential threat to containment integrity.

4. Zircaloy itself starts to melt around 1700°C; the sequence of melting may be as illustrated in Figure 6.1. Molten droplets and rivulets of eutectic are formed (rather like wax running down a candle; Figure 6.1a). They solidify in the lower, cooler regions of the core, causing further blockage, which exacerbates the lack of cooling (Figure 6.1b). The solidified material forms a crucible. With the cladding around them gone, the fuel pellet stacks are unstable. Any transient (like the starting of the primary pumps in the TMI-2 accident) can cause a redistribution of this material with the pellets falling into the crucible to form a debris bed. This material is still generating heat and there will be a tendency for it to melt and move down through the core, growing in volume as it does so (Figure 6.1d)

5. The mass of molten material will eventually reach the bottom core support plate and will be held there for a period of time until that core plate also fails

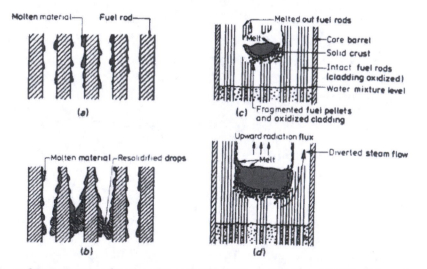

Figure 6.1: Sequence of core melting. Initial stages: (a) Molten droplets and rivulets beginning to flow down intact fuel rods; (b) formation of local blockage in colder regions of fuel rods and formation and growth of a molten pool; (c) formation of a small molten pool; (d) radial and axial growth of the pool.

and the core debris then has access to the lower plenum of the reactor pressure vessel.

6.2.2 Challenges to the Reactor Pressure Vessel

The reactor pressure vessel represents the second containment barrier or line of defense. It consists of a massively thick ferritic steel structure. What are the possible challenges to the integrity of this vessel? First, it is designed, constructed, and inspected to the highest quality standards. Failure of the vessel due to internal or external loadings *within* the design basis is considered incredible. However, various failure modes or mechanisms can be postulated in severe beyond-design-basis accidents. Thus the vessel might fail due to

- gross overpressurization
- displacement or damage from the support structures
- creep failure due to overheating or vessel wall thinning
- shock loadings due to internal fuel-coolant interactions or hydrogen explosions

Overpressurization could occur as a result of the reactors failing to trip or shut down in response to an operational transient. An example would be some fault that removes the coupling of the reactor to the heat sink, which unless the reactor is shut down will lead to a rapid rise of primary circuit pressure. In practice initially the negative temperature coefficient, followed by the lifting of the relief valves and subsequent voiding of the core, limits the reactor power and ultimately terminates the fission reaction. As a result the primary pressure peaks well under that which might cause failure of the vessel (~400 bars).

Displacement or removal of the vessel from its supports could occur due to an earthquake of exceptional magnitude or, alternatively, as a result of internal shock loadings from within the vessel itself (see below).

As indicated above, the process of core degradation and melting could result at some point in molten fuel or other materials entering the lower plenum of the reactor vessel. In the TMI-2 accident some 20 tonnes of molten fuel ended up in this location.

What the consequences are depends on how the core support plate fails and when. The lower part of the vessel may still contain a pool of water notwithstanding the high temperatures existing in the upper part of the vessel. If the mass of molten material above the lower core plate jets into the pool of water, a "steam explosion" may occur and may damage the vessel. Such events are discussed in Section 6.3. Alternatively the jet of heavy molten fuel may penetrate to the vessel wall and result in rapid heating of the wall. Wall thinning will result, and if the vessel is still at high pressure, plastic collapse may occur. This may happen quite rapidly—within minutes. Rather than forming a jet, the molten fuel may enter the plenum at the periphery of the core support plate or through the baffle plate and pour down the side of the lower head as in the TMI-2 accident. This more gentle process may not provoke a fuel-coolant interaction. However, if a pool of molten material does form in the lower head, overheating, wall thinning, and ultimately creep failure may result. Failure of the penetrations for the in-core instrumentation may also occur. Evidence from inspection of the lower head of the TMI-2 vessel, however, suggests that some considerable cooling was available via cracks in the fuel debris and via the gap between the debris and the vessel wall. Although some damage was observed of the vessel wall and the penetrations, no failure of the pressure boundary occurred.

If the failure of the bottom core support plate results in jets of molten fuel entering a pool of water in the lower plenum, a fuel-coolant interaction "steam explosion" may result. This is particularly the case if the pressure in the primary system is low. This process may cause damage inside and outside the vessel. In the worst case the vessel may be lifted off its supports and/or the head of the vessel may be blown off, damaging the containment. To ensure that the containment is not damaged, it is important to show either that missiles with sufficient kinetic energy are not formed or alternatively that the containment structures can accommodate the missile without damage to the containment function. Typically if the fuel-coolant interaction produced an explosive energy input of 1 GJ, then a considerable portion of the kinetic energy of the molten core slug projected upward through the vessel is absorbed by plastic deformation of the internal core structures and stretching of the vessel bolts. Perhaps only 5–10% of the total energy will be imparted to the upper head. This process occurred during the course of the SL-1 accident (Section 5.2.1).

Finally, the intense, and possibly explosive, interaction of molten fuel with water will cause the fuel to be dispersed as small particles. These could form a *debris bed* in the bottom of the vessel. Depending on the size of the particles and the ability of the operators to continue to feed water to the vessel, it may be possible to cool this debris bed over a long period. This would terminate this class of accident without release of radioactive material into the containment. If it is not possible to cool the fuel debris, the bottom head will fail, releasing the molten fuel into the cavity in which the vessel sits.

6.2.3   Challenges to the Reactor Containment

The final "defense in depth" barrier to the release of radioactive materials to the environment is the containment building or system itself. The TMI-2 accident demonstrated the importance of the reactor containment in converting a very severe accident in the reactor itself into one that had very little public health impact. There has been much study of the integrity of containments, including experimental research under simulated accident conditions, particularly for PWRs.

The main forms of reactor containment that have been employed are as follows:

- Large prestressed or reinforced concrete shells that are designed to withstand internal pressures of 3–4 bars above atmospheric.
- Spherical steel vessels (as used in German reactors) that are similar in concept to the concrete vessels and withstand about the same pressure.
- Steel or concrete vessels in which ice is used to condense any steam released from the reactor system (so-called ice-condenser plants). Here the design pressure can be lower, but such concepts are less popular than they used to be since the ice-condenser is of little help in containing, say, a hydrogen explosion.
- Pressure-suppression containments in which the system is arranged so that any steam escaping from the reactor circuit will bypass through vent tubes into a pool of cold water where it is condensed.

Advanced containment systems often involve a double wall containment with a steel or prestressed concrete inner wall and a reinforced concrete outer wall. A subatmospheric pressure is maintained in the interwall space. Alternatively it is possible to combine a pressure-suppression system inside a conventional dry-well containment.

The likelihood of an accident's leading to a breach in the containment is low. As exemplified by the case of TMI-2, in the majority of severe accidents the containment will fulfill its function. Challenges, however, can come from overheating or overpressurization, hydrogen explosion, or missile impacts. These could result in structural failure or damage to the liner or a penetration resulting in a high rate of leakage. In addition, failure to isolate the containment during an accident could allow the transfer of radioactivity to other parts of the plant or to the environment. The timing of any failure is also relevant. The longer the containment remains intact, the greater the opportunity to take action to protect the public from any release.

In the previous section we saw that if molten fuel reaches the reactor vessel lower head, then this may fail. If this failure occurs rapidly with the primary system still at high pressure, the molten fuel will be ejected into the reactor cavity and from there it can move into the containment building. Rapid heating and pressurization of the containment will result from

- molten debris particles heating the containment atmosphere
- chemical reactions between the debris and water-steam leading to additional heating
- hydrogen, produced by chemical reactions, burning or detonating

Tests have been carried out at the Sandia National Laboratories using one-tenth-scale containments modeling the geometric details of actual nuclear power plants. Iron/aluminium/chromium thermite was used to simulate the molten core. This was ejected by high-pressure steam from the scale pressure vessel bottom head. Pressures and temperatures inside the containment were measured. Water that might flood the reactor cavity or be on the containment basemat was present in some experiments.

The results demonstrated that heating of the containment is less if the debris is contained below the main operating deck. Water in the reactor cavity reduces the pressurization due to the steam released by quenching the melt. Water on the basemat has little or no effect. Hydrogen produced by the cavity interaction and dispersal processes burned with a diffusion flame in the upper dome and contributed about half to the pressurization. Any preexisting hydrogen burned slowly and had little effect. These experiments and modeling calculations suggest that a wide range of variables influence containment heating

- Primary circuit pressure prior to vessel failure
- Mass of molten core in pressure vessel lower head
- Temperature and composition of the molten core
- Particle-size distribution and entrapment of debris
- Impingement onto surfaces and freezing
- Chemical reaction of debris with steam
- Quenching by reactor cavity water
- Formation, transport, and burning of hydrogen

From a variety of studies it can be judged that containment integrity will be at risk only if a large fraction of the core is ejected and a hydrogen detonation occurs. In these circumstances, the containment may be subjected to continuous static pressures of 10–15 bars as well as a transient pressure pulse. However, a major mitigating feature would be depressurization of the primary circuit prior to molten core ejection so as to limit debris dispersion. Depressurization to 15–25 bars is effective in this way (see Section 6.2.4).

Even if the containment survives the early containment heating and pressurization, there are still challenges to its integrity that occur later. These include

- hydrogen combustion
- gradual overpressurization
- basemat melt-through

Hydrogen formation and combustion are described in Section 6.3.3. Slow overpressurization may occur if there is no heat removal or venting of the containment. Moreover, interaction of the core debris with concrete produces copious amounts of noncondensible gases such as carbon monoxide and carbon dioxide to add to the pressurization.

The interaction between the molten core and the concrete depends on a number of factors including the presence of water. If the cavity is initially dry and the core debris forms a deep bed, then extensive interaction may occur. Flooding the debris may effectively cool the melt. Some comments about the coolability of debris beds are made in Section 6.3.2.

There remains a finite possibility that the molten core materials may attack the containment basemat. This scenario is discussed in Section 6.3.4. Complete melt-through of the basemat would take several days, but the consequences of failure are relatively small compared with the failure of the containment above ground.

Finally, the integrity of the containment building may be compromised by the failure to isolate the building. As explained in Chapter 4, once the emergency core cooling system in the PWR is initiated, the containment is isolated. In practice, a number of essential services must still be provided to the reactor (emergency feedwater, etc.), and this provides a number of routes for release from the containment. For example, rupture of the decay heat removal system outside the containment, coupled with failure of an isolation valve, could give a route out from the containment that bypasses the building itself. Proper securing of personnel and equipment airlocks is also essential. Two specific accident sequences are important in this context:

The so-called interfacing systems LOCA in which important check (or nonreturn) valves fail and low-pressure piping connected to the reactor coolant system fails *outside* the containment. This provides a bypass to the enviroment.

Failure of steam generator tubes during the course of an accident again may permit bypass to the enviroment via the secondary side steam relief valves.

6.2.4 Mitigating the Consequences of Severe Accidents

The next generation of nuclear power plants will incorporate design features that will eliminate or reduce the challenges to the various containment barriers or mitigate the consequences of failure. One example is the design features included on the European pressurized water reactor (EPR). The EPR design includes:

- The elimination of situations where the degradation of the core occurs with the primary circuit still at high pressure. This is achieved by high-reliability secondary side–decay heat removal systems but also by means of rapid depressurization via the pressurizer relief valves.
- The elimination of direct containment heating via the depressurization facility.
- The limitation of the containment pressure increase using a dedicated spray heat removal system that can subcool the water and return the pressure to atmospheric. The containment design pressure of 7.5 bars allows 12–24 hours after the accident before it is necessary to use the spray system.
- The provision of a double-wall containment with collection of all leaks in the interwall space where a lower pressure is maintained.
- The prevention of hydrogen explosions by reducing the hydrogen concentration using catalytic recombiners together with selectively placed igniters.
- Accommodation of the consequences of an instantaneous full cross section rupture of the reactor pressure vessel at a pressure of 20 bars via careful design of the layout.
- Provision to cope with molten fuel coming from a failed pressure vessel lower head, first, without a "steam explosion" and, second, preventing interaction with the containment concrete.

This is accomplished as shown in Figure 6.2 by connecting the reactor cavity

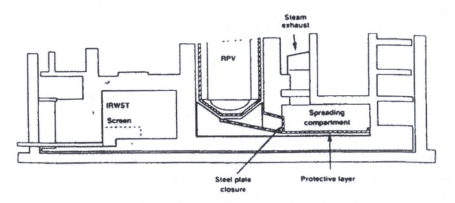

Figure 6.2: EPR cone spreading feature.

to a dedicated molten core spreading chamber via a refractory lined melt discharge channel. The spreading chamber has a large area (150 m^2) and is normally sealed from the reactor cavity by a steel plate. This plate resists melt-through for a limited time in order to accumulate the molten fuel in the cavity. The spreading compartment is connected via pipes to the refueling water storage tank in the containment. These pipes are normally closed by fusible plugs. This ensures that the water floods the spreading chamber only *after* the melt has been spread over the area of the chamber.

6.3 SPECIFIC PHENOMENA RELATING TO SEVERE ACCIDENTS

In the previous section reference is made to a number of specific phenomena that can directly influence the course of a severe accident. Rather than interrupt the discussion in that section of the effectiveness of the three containment barriers to cover these phenomena in detail, it was more convenient to deal with these topics in a separate section. Thus, this section covers

- fuel-coolant interactions—"steam explosions"
- debris beds and their cooling
- hydrogen formation—burning and explosions
- containment basemat melt-through and failure

6.3.1   Fuel–Coolant Interactions: "Steam Explosions"

When a liquid comes into contact with another liquid and the first liquid is at a temperature much greater than the boiling point of the second liquid, rapid vaporization of the second liquid may occur as the first liquid cools. Under some circumstances, this rapid vaporization may cause a detonation. Such detonations have been observed in metal foundries where vats of molten metal have been accidentally poured into vessels of water, or vice versa. They may also occur if room-temperature water is brought into contact with liquid natural gas; in this case, the detonation may be followed by a fire as the gas cloud burns. The potential for an energetic interaction between molten uranium fuel and the water coolant may also exist if molten fuel is jetted into water. This can occur as

- molten fuel is ejected into the coolant when the cladding fails during a severe power excursion (cf. Chernobyl).

- the lower core support plate fails and molten fuel is jetted into a pool of water in the vessel lower head
- the lower head of the pressure vessel fails and molten fuel falls into a water-filled reactor cavity

The circumstances arising in a fuel-coolant interaction and leading to a vapor explosion are illustrated in Figure 6.3. The molten fuel is initially above the pool of coolant (Figure 6.3*a*) and then falls into it (Figure 6.3*b*), giving rise to coarse mixing between the fuel and the coolant with a dispersion of large elements of the molten fuel as illustrated. These elements might be 1 cm in diameter. They transfer heat relatively slowly to the water, since a thin vapor film forms around them and insulates them from the water coolant. The third stage is that of triggering a shock wave. This is often postulated to occur at the surface of the vessel (Figure 6.3*c*) and might be caused by a small, localized vapor explosion or impact. This shock wave then passes through the coarse fuel-coolant mixture and breaks up the fuel into small elements, which may transfer their stored en-

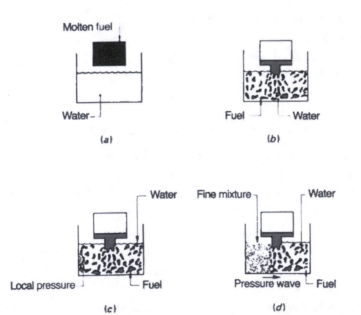

Figure 6.3 Stages of a steam explosion. (*a*) Initial condition; molten fuel and coolant separate. (*b*) Stage 1, coarse mixing; slow heat transfer, no pressure increase. (*c*) Stage 2, trigger process; local pressure, e.g. from impact or entrapment. (*d*) Stage 3, propagation; pressure wave fragments fuel very rapidly; heat transfer from fine fragments very rapid. (Gittus et al. 1982.)

ergy rapidly to the coolant. This energy release strengthens the shock wave, which continues to propagate through the mixture in an explosive manner (Figure 6.3d).

The energy stored by the molten fuel on release into the coolant pool is partly converted to energy in the shock wave. The extent of this conversion is obviously very important in considering the effects of the resultant shock wave on the reactor system. Experimental studies indicate that the efficiency of conversion from the stored energy in the fuel to the energy within the explosion is about 1.5%.

This would result in an explosion of roughly 1 GJ (or 200 kg TNT equivalent) if all the fuel in a PWR, say, reacted simultaneously.

There is still considerable discussion about the precise mechanism by which the shock wave propagates through the fuel-coolant mixture. One theory suggests that associated with the shock, there is spontaneous formation of vapor bubbles, giving rise to rapid transfer of energy from the fuel to the coolant. Another theory suggests that in the shock itself the mechanism of heat transfer is quite different, with the fuel being shredded to small elements by the shear forces in the shock and these elements transferring their energy rapidly to the coolant behind the shock. As we have discussed earlier, a high-pressure impulse resulting from a steam explosion is transmitted into the coolant pond. This accelerates a slug of coolant, which impacts the upper head of the vessel and might induce failure. The influence of steam explosions in reactor systems is still a subject of debate, and no final judgments can be made at this time.

6.3.2   Debris Beds and Their Cooling

As we saw in the previous section, there are a number of circumstances in which beds of fuel debris may be formed, initially submerged in a pool of coolant. If such beds can be effectively cooled, remelting is avoided and damage to the vessel or the cavity contained in the bed may be prevented. In recent years, and particularly since the accident at Three Mile Island, much attention has been given to the coolability of such beds.

The cooling of beds of debris is a highly complex process and is strongly affected by such variables as the bed particle size, the means of access of the coolant to the bed, the bed depth, and the system pressure. Some mechanisms for debris bed cooling, illustrated in Figure 6.4, are as follows:

1. *Once-through flow through the bed.* Here it is assumed that the liquid is able to reach the bottom of the bed and is then induced to flow into the bed under the action of natural or forced circulation. Natural circulation would be caused by the difference in density of the coolant inside the bed and outside the bed. This is the same kind of circulation that occurs in some forms of steam-generating boilers. Alternatively, the debris bed may be in a region of the reactor over which a pressure drop occurs in the circulation liquid, and this pressure drop would force liquid through the bed. As illustrated in Figure 6.4*a*, the first phase is for the heat generated in the bed (from decay heat of the fission products trapped in the bed) to heat the liquid to its boiling point. Then, as the flow passes through the bed, the liquid is evaporated and ultimately converted totally to vapor. From this point on, the temperature rises rapidly with distance up the bed, and if the circulation is too low or the bed too deep, the particles may reach a temperature at which they begin to fuse together. This clearly represents a limit to this form of cooling.

2. *Cooling of closed deep beds.* Here, as illustrated in Figure 6.4*b*, the liquid can only enter from the top of the bed. The liquid trickles into the bed, cooling it and generating vapor, which must escape in the direction opposite to that of the liquid flowing in. This causes a flooding phenomenon of the type we discussed in Chapter 2, with the vapor resisting and limiting the entry of liquid at the top of the bed. This may mean that only the upper part of the bed is cooled and the lower part may become overheated. This limitation is more severe the smaller the particle size in the bed. Again, drying out and fusing of the lower part of the bed is the limit on cooling in this situation.

3. *Shallow-bed cooling.* If there is a shallow bed of particulate material on the bottom of the containment and this is covered by a liquid layer, then "chimneys" may be formed in the layer (Figure 6.4*c*) through which the vapor may escape, the liquid passing into the bed by capillary action through the particulate layer between the chimneys. This is an efficient way of cooling but can only be applied over a limited range of conditions.

Experiments and calculations show that in case 2, for a 1-m-deep bed, a heat dissipation rate of 750 kW/m^3 may be achieved if the particles are 4 mm in diameter in a pool of water at 1 bar (atmospheric pressure). However, the maximum dissipation rate before dryout and fusion of a bed composed of particles of 0.1 mm diameter would be only 20 kW/m^3. Thus, the effectiveness of the debris bed cooling can be estimated accurately only if the particle size of the bed is known. Although a better understanding of the mechanisms of debris bed cooling is now beginnng to emerge, the main difficulty of predicting the particle size that might result in different phases of the accident is still to be resolved. A typical debris bed might have a power generation rate (from fission product decay) of 1000 kW/m^3 some 3 h after initiation of the accident—about the time at which one might expect meltdown in a PWR. This power could be

Liquid pool with vapour bubbles generated in bed

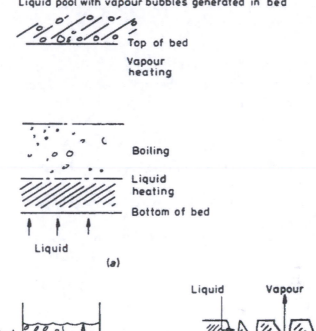

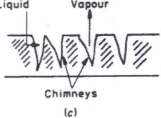

Figure 6.4: Debris bed cooling.

dissipated in a bed 0.5 m thick provided the particle size was greater than 2 mm. These calculations are for a PWR, but a similar picture is obtained for the fast reactor, since its increased fuel rating (and hence fission product decay heating) is offset by the increase in latent heat of vaporization of sodium compared with that of water.

6.3.3   Hydrogen Formation: Burning and Explosions

Hydrogen can be formed at various stages of a severe accident as a result of chemical accidents between steam and various metals. This hydrogen can burn or detonate, hazarding the containment systems.

The most important contributor to the hydrogen formation process is the oxidation of the zirconium cladding of the fuel:

$$Zr + 2H_2O = Zr\ O_2 + 2H_2$$

The reaction is exothermic adding to the decay heat. The extent of the chemical reaction is determined by a number of factors, including the access of steam to unreacted metal and the geometry of the core debris. Other materials that react include chromium and iron and even uranium dioxide.

Hydrogen may be formed at various phases of the accident:

1. When the initial heat-up occurs; perhaps 20–40% of the cladding may react in the first 10 or 20 minutes
2. When further water from the ECCS system or reactor coolant pumps contacts the hot debris
3. When molten debris jets or falls into the vessel lower head and vaporizes water to steam, which then has access to relatively undamaged fuel in the core above
4. When the pressure lower head fails and the molten debris attacks the concrete of the vessel cavity and containment

In the case of a large loss-of-coolant accident (LOCA), the hydrogen may be released to the containment as it is formed. Conversely, where the primary circuit remains intact, the hydrogen release may occur at the time of lower head failure.

Hydrogen can react with the oxygen within the containment in one of two ways. The first way is by deflagration, or a diffusion flame in which the unburned gas is heated by conduction to a temperature sufficiently high for a chemical reaction. Whether a combustion reaction takes place depends on reaching the minimum concentration of the hydrogen, i.e., 4–9% by volume. While diffusion flames and slow deflagrations add to the heating load and therefore the pressurization of the containment, they do not represent a serious threat to the integrity of most designs. Such a deflagration occurred during the TM1-2 accident.

In the second way, in a detonation, the unburned gas is heated by compression in a shock wave. Initiation can come from a spark or other high-energy source. The consequences of a detonation depend on the concentration of the hydrogen (the higher the concentration the higher the combustion pressure) and the geometry of the containment internals.

One means of controlling hydrogen is to have an inert atmosphere (nitrogen) in the containment. This is used on some reactor designs, particularly those of BWRs, but has operational disadvantages. Other techniques include catalytic recombiners (which react hydrogen and oxygen to form steam) and igniters (which deliberately ignite the hydrogen at the lower mixture concentration) installed at various locations within the containment.

6.3.4   Containment Basemat Melt-Through and Failure

If it is not possible to cool the debris bed within the containment building, the debris begins to react with the concrete floor of the building and penetrates this and also the bedrock on which the reactor is built. This gradual downward penetration of the molten pool has colloquially been referred to as the "China Syndrome," it being imagined that the pool could ultimately penetrate through to the other side of the earth, which in the case of the United States is imagined to be China. Actually, this imagined situation is impossible: the pool would miss China by a long way and could only pass outward from the center of the earth if gravity mysteriously became negative. However, penetration of the molten material is limited.

Figure 6.5 shows an overall diagram for the containment for a PWR. Turland and Peckover (1979) calculated the behavior of a molten pool arising from a 3-GW(t) reactor core. There are two extreme situations.

First, if the melt consists mainly of oxide, it is likely to be miscible with the base concrete and rock. A molten pool would be formed of limited depth (around 3 m) and with a diameter of about 13 m (Figure 6.5). This pool will remain for a period of up to several years. Figure 6.5 illustrates the situation after 1 year and shows the temperature profile in the rock-concrete around the pool. The heat generated by fission product decay within the pool is dissipated into the surrounding rock due to the temperature gradients illustrated.

Second, if in the melting process molten steel is produced, this may dissolve fission products from the fuel. If this molten steel is oxidized, the melt pool will be miscible with the concrete-rock base and a pool such as that illustrated in Figure 6.6 will be formed. If the steel is not oxidized, the steel-fission product solution will not be miscible with molten fuel and concrete-rock and will itself penetrate the base rock much farther. Calculations by Turland and Peckover (1978) are illustrated in Figure 6.7. It shows that a molten metal, immiscible pool of this type could penetrate to a maximum depth of about 14 m.

The two melt pools illustrated in Figures 6.6 and 6.7 are drawn in scale in the diagram of the containment shown in Figure 6.5.

It is noteworthy that the interaction between the molten fuel and the concrete-rock will result in the release of significant amounts of vapor and gas as a

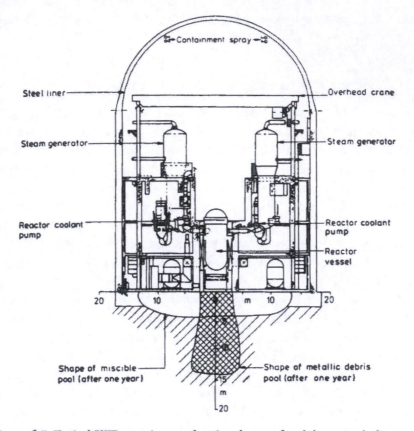

Figure 6.5: Typical PWR containment showing shapes of meltdown pool after 1 year.

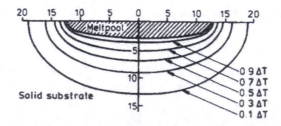

Figure 6.6: Shape after 1 year of an axisymmetric miscible pool for core debris from 3-GW(t) core (gas agitation neglected). The substrate isotherms are labeled with their temperature excess above ambient.

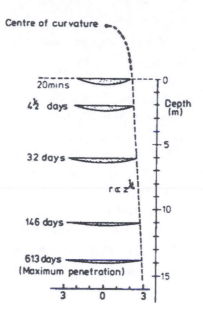

Figure 6.7: Descent of lens-shape pool (volume, 3m³). $\lambda Q_0 = 100$ MW; k=2 W/mK.

result of the chemical reaction. This may result in pressurization of the containment building over a long period of time, particularly if no cooling is available.

As the fission products in the pool of material decay, the molten fuel gradually solidifies. Calculations indicate that the pool of molten material under the reactor might reach a maximum size equivalent to a hemisphere about 27 m in diameter. Because a considerable amount of concrete is mixed with the fuel, it has been suggested that the final form of the solidified mass is likely to be a glasslike substance that would immobilize the fission products and limit their subsequent migration.

As we have seen above, even the worst case of fuel meltdown and failure to cool would lead to an acceptable situation *provided there is no failure of the containment.*

6.4   SEVERE ACCIDENTS IN OTHER REACTOR TYPES

The sequence of events outlined in Section 6.2 applies to a PWR; the situation with regard to other reactor types can be summarized as follows:

Boiling-Water Reactor (BWR). The situation is very similar to that in the PWR regarding sequences of core meltdown, fuel-water interaction, and ultimate disposition of the molten fuel pool.

CANDU. The melting sequence is not considered to be very likely because of the large pool of moderator heavy water through which the individual fuel channels pass. Analysis of the heat transfer events following a loss-of-coolant accident and failure of the emergency core cooling system has indicated that significant fuel melting would not occur and provided the means of extracting heat from the moderator were still intact, the accident would be controlled. However, should a single-pressure tube fail and the moderator become pressurized as a result of the release of high-pressure steam into it, the moderator could be expelled and its cooling effectiveness for the other channels removed. That this is at least a remote possibility is indicated by the accident at Lucens, described in Chapter 5. If the moderator was expelled, fuel melting would proceed in the same way as for the other water reactors; again, this event could be contained provided there were no steam explosions or other events that disrupted the containment.

Magnox Reactor. The inherent basic safety features of the Magnox reactor (the fact that the graphite itself may absorb a great deal of heat and that decay heat removal can be maintained even if the reactor is depressurized) have led to the view that a full core meltdown is not credible. However, some studies have been done on the effects of meltdown of single channels, specifically those with the highest rating. That such single-channel events are credible is borne out by the accidents in this type of reactor discussed in Chapter 5. In the Magnox reactor systems, such events can lead to small releases of activity since the reactors do not have the hermetic containment that is provided for water reactors.

Advanced Gas-Cooled Reactors (AGRs). Full meltdown accidents are not considered credible for this type of reactor for much the same reasons as mentioned above for the Magnox reactors. Furthermore, with AGRs, much higher fuel temperatures can be sustained before fuel damage since the fuel is in the oxide form and clad in stainless steel (in a Magnox reactor the fuel is in the form of uranium metal clad in magnesium alloy). Tests in the Windscale prototype AGR showed that the fuel temperature can approach to within 50°C of the melting point of steel without clad meltdown and significant fuel damage. However, single-channel fuel melting due to local blockage effects, or due to the dropping of a fuel stringer during the refueling operation, is still considered possible and is taken account of in the design. As explained in Chapter 4, the rise of temperature following a loss-of-coolant accident in an AGR is very slow indeed compared with that in a PWR or a BWR. This means that there is time to take alternative actions, even if off-site power is lost and the local power supplies feeding the emergency circulators fail to operate immediately. It is interesting to compare the situations in an AGR and PWR; in the AGR the consequences of a fuel meltdown would be more serious since it does not have hermetic containment; on the other hand, the probability of a meltdown is even smaller than in the case of the PWR.

Liquid Metal–Cooled Fast Reactors. The very high fuel ratings in fast reactors have led to much interest in the possibility of core meltdown and its consequences. One accident scenario is that of failure of all the primary sodium coolant pumps and complete failure of the reactor shutdown system. As the sodium reaches its boiling point in the channels of maximum rating, sodium boiling and voiding occur, and this has a net positive reactivity effect on the reactor, which accelerates the heating. Melting of the fuel and cladding occurs in about one second after sodium voids are formed in a particular fuel assembly. In the area of that assembly there is a complex mixture of liquid fuel, sodium vapor, liquid steel, fuel fragments, fission gas, and steel vapor. If the fuel channel walls melt, adjacent channels may also be damaged and melted.

Calculations of the consequences of these events are highly complex because of the coupling between the nuclear reactions, the heat transfer processes, and the fluid flow processes. Two different outcomes are possible, depending on such things as the reactor design and reactor state at the beginning of the accident:

1. If, during the meltdown, a large fraction of the original fuel has managed to remain within the active core region, an extremely large increase in reactivity occurs and the fuel is actually blown apart and dispersed by the fission product gases in the interstices of the fuel pellets. The dispersal of the fuel terminates the nuclear reaction, though the resultant shock wave may damage the reactor structure and breach the containment.

2. If the fuel inventory has been reduced to about half the original amount by gradual leakage, or if large quantities of blanket materials have diluted the fuel, a severe power excursion will not occur. The molten fuel will fall to the

bottom of the reactor and the sequence of events will be similar to that de-scribed above for the PWR, including the possibility of a vapor explosion due to interaction between the molten fuel and the liquid sodium still in the vessel. The possibility of this form of accident has drawn great attention to the reliability of shutdown systems in fast reactors; one possible design ap-proach is to arrange the core structures so that an excessive increase in core temperature causing its thermal expansion will trigger an automatic shut-down of the reactor. Combined with the fact that the decay heat can be re-moved by natural circulation to air-cooled heat exchangers and the enormous heat capacity of the sodium coolant, this inherent shutdown sys-tem would give the fast reactor system a "walkaway" safety capability that is not available in other reactors, which depend on the operation of active sys-tems demanding operator actions and/or totally reliable power supplies.

Clearly the attention given to core meltdown accidents varies from reactor to re-actor and depends on the assigned credibility for such accidents. In general, the objective is to bring down the likelihood of an accident and in particular its public consequences to a minimal level.

6.5   FISSION PRODUCT DISPERSION FOLLOWING CONTAINMENT FAILURE

Should the containment fail, fission products will be released into the atmos-phere. There is much discussion about the extent to which this would happen. The gaseous fission products are usually assumed to be released completely and other volatile fission products such as caesium and iodine are assumed to be partly released. Other fission products are released in very small quantities and do not usually contribute significantly to the calculated hazard. Typical fractions of caesium and iodine assumed to be released are around 10%. For the less volatile fission products, fractions around 1% are often assumed, on the basis of experimental studies of fission product retention. The ultimate disper-sal of the fission products from such a release is calculated by using computer codes and depends greatly on the weather conditions at the time of release.

REFERENCES

Baumgartl, B.J., and F. Bouteille (1994). "The European Pressurized Water Reactor (EPR): An Advanced PWR." *Revue Generale Nucleaire,* November–December, 478–483.

Gittus, J.H., et al. (1982). "PWR Degraded Core Analysis." Report ND-R-610(S), U.K. Atomic Energy Authority.

Hennies, H.H. (1993). "Research and Development to Improve Containment for the Next Generation of Pressurized Water Reactor Plants." *Interdisciplinary Science Reviews* 18 (3): 243–51.

Pilch, M., Allen, M., and T. Blanchat (1994). "Can Containment Buildings Take the Heat?" *Atom* 434 (June–July).

Prior, R. (1992). *Severe Accident Progression. Core and Reactor Systems Phenomena.* Severe Accidents and Accident Management in Light Water Reactors, March 23–27, Lyon, France.

The Three Mile Island Reactor Pressure Vessel Investigation Project: Achievement and Results. (1993). Proceedings of an Open Forum Sponsored by the OECD/NEA and the US NRC, Boston, October 20–22.

Turland, B.D., and R.F. Peckover (1978). "Melting Front Phenomena." Report CLM-P-564, U.K. Atomic Energy Authority..

————(1979). "Melting Front Phenomena." *Eur. Appl. Res. Rept.* 1 (6): 185–201.

EXAMPLES AND PROBLEMS

1 Total decay heat from a reactor

Example: The total amount of decay heat that can be generated from a reactor core is finite; eventually, all the fission products decay to a nonradjoactive state and the energy that is released in this long-term process is fixed and can be calculated by estimating the energy release from the decay of each relevant fission product and summing the energy released from all fission products. As an example of this process, calculate the total decay heat released from 1 kg of iodine-I 31 that is present in the reactor at shutdown. Iodine decays to xenon-131 by the reaction

$$I\text{-}131 \rightarrow Xe\text{-}131 + \beta + \gamma$$

each atom that decays releases 0.57 MeV (9.12×10^{-14} joules) of energy. Assuming a half-life of 8 days for I-131, what fraction of this energy is released in the first 30 days of decay?

Solution: The number of atoms (N) of I-131 in 1 kg is given by

$$N = \frac{\text{number of atoms per kg mole}}{\text{atomic weight}}$$

$$= \frac{6.022 \times 10^{26}}{131}$$

$$= 4.597 \times 10^{24} \text{atoms/kg}$$

$$\text{Total energy released} = 4.597 \times 10^{24} \times 9.12 \times 10^{-14} \text{ J}$$

$$= 4.192 \times 10^{11} \text{ J} = 0.419 \text{ terrajoules (TJ)}$$

The number N_t of atoms of I-131 remaining after t days is given by

$$N_t N \exp(-\lambda t)$$

where λ is the decay constant, in reciprocal days. From the substitution

$$0.5N = N \exp(-8\lambda)$$

we have

$$-8\lambda = \ln 0.5$$
$$\lambda = 0.8664$$

The number of atoms remaining after 30 days is given by

$$N_{30} = N \exp(-0.08664 \times 30)$$
$$= 0.0743N$$

Thus, $1 - 0.0743 = 0.926$ of the energy released by decay of the iodine-131 will have been released in the first 30 days.

Problem: Calculations reveal that the total amount of decay heat released from the core of a 1000-MW(e) reactor following shutdown is around 100 TW. If all this heat is released into a boiling-water pool, what would be the total mass of water evaporated from the pool if the latent heat of evaporation is 2257 J/kg? If the heat were released and absorbed in melting the concrete surrounding the reactor, what mass of concrete would be melted (assuming no heat loss to the environment)? If the molten concrete formed a hemispherical pool, what would be the radius of this pool? (Assume that the latent heat of melting of concrete is 1000 J/kg and that the density of the molten concrete is 2000 kg/m^3)

2 Formation and cooling of debris beds

Example: Following a LOCA and subsequent failure of the ECCS system in a PWR, the reactor core partly melts and the molten material falls into the water pool at the bottom of the reactor vessel, forming a 0.75-m-deep submerged particle bed with a porosity (fraction of total volume of bed that is free of the solid phase) of 0.4. The decay heat release rate from the bed is 1000 kW/m^3 of bed after 3 h.

Use the data shown in the following figure (from Gittus et al., 1983) to estimate the minimum particle size that would be needed for cooling of the bed by ingress of water into the bed, without dryout of the bed:

Solution: Interpolating from the figure of Gittus et al., we can plot the dryout heat release rate for a bed of 0.75 m depth as a function of particle size as follows:

Extrapolating this curve slightly, we see that a minimum particle size of about 0.40 mm would be required if the stated decay heat (1000 kW/m^3) is to be accommodated without dryout.

Problem: Using the data of Table 2.2, estimate the heat release rate from the particle bed at 10 and 100 h (assuming 1000 kW/m^3 at 3 h). Estimate the minimum particle size that would be required to allow heat dissipation without dryout for beds of 0.75 m depth formed at 10 h and 100 h after shutdown.

3 Steam explosions

Example: In a severe accident in a PWR, 50 metric tons (50,000) kg) of molten core

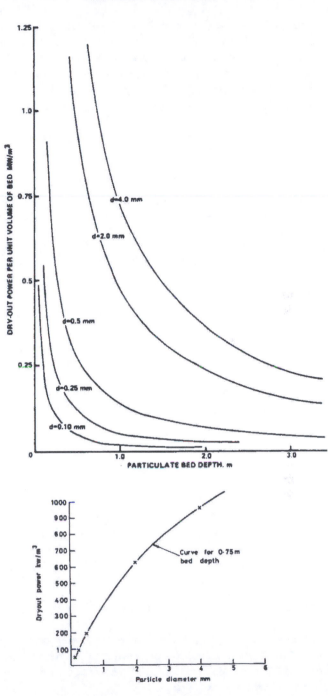

Problem 2 *Formation and cooling of debris beds*

material at 3000 K is released into a pool of water remaining at the bottom of the reactor pressure vessel. A steam explosion occurs, releasing 3% of the original thermal energy of the fuel, and the energy of the explosion is transmitted to a 10-ton slug of water that rises up the vessel, hitting the top of the vessel. At this stage the vessel is unrestrained and weighs (with its contents) 500 tons. Calculate the height that the vessel rises as a result of the impact with the water slug. Assume that the thermal energy of the fuel is 1.5 GJ per ton.

Solution: The amount of energy released by the explosion is given by:

$$E = 50 \times 0.03 \times 1.5 \times 10^9 \text{ J}$$
$$E = 2.25 \times 10^9 \text{ J} = 2.25 \text{ GJ}$$

If this energy is transmitted into the kinetic energy of a 10-ton water slug. we may calculate the velocity V_s of the water slug, since the kinetic energy of the slug is given by $1/2 \, m \, V_s^2$ where m is its mass. Thus

$$\tfrac{1}{2} \, mV_s^2 = 2.25 \times 10^9 \text{ J}$$

and

$$V_s = \left(\frac{2.25 \times 10^9 \times 2}{10,000} \right)^{\tfrac{1}{2}} = 670.8 \ m/s$$

After the impact with the vessel the vessel begins to rise with a velocity V_v. The principles of conservation of momentum apply and we may write

$$mV_s = MV_v$$

where M is the mass of the vessel. Thus:

$$V_v \frac{m}{M} V_s = \frac{10}{500} \times 670.8$$
$$= 13.42 \text{ m/s}$$

The kinetic energy of the vessel after impact is given by

$$\text{Kinetic energy} = \frac{1}{2} MV_v^2 = \frac{1}{2} \times 5 \times 10^5 \times 13.42^2$$
$$= 4.50 \times 10^7 \text{ J}$$
$$= 45.0 \text{ MJ}$$

Note that a large proportion of the original energy of the water slug has been lost in the impact.

The initial kinetic energy of the vessel is converted to potential energy as the vessel rises from its original position. Suppose the vessel rises to a height b before coming to rest. The potential energy gained at this height relative to the original position is Mgb, where g is the acceleration due to gravity.

Thus:

$$Mgb = \frac{1}{2}MV_v^2 = 4.5 \times 10^7 \text{ J}$$

$$b = \frac{4.5 \times 10^7}{Mg} = \frac{4.5 \times 10^7}{5 \times 10^5 \times 9.81}$$

$$= 9.17 \text{ m}$$

A rise of 9.17 m would not normally be sufficient to bring the vessel into contact with the containment, but it would probably lead to its hitting the missile shield above the vessel, depending on the system design.

Problem: Calculations reported by Gittus et al. (1982) suggest that a 0.4-m-diameter missile of 850 kg traveling at 300 m/s would penetrate both the missile shield and the containment, giving rise to containment failure and partial release of fission products to the environment. In the steam explosion described in the example, suppose the impact between the water slug and the upper part of the vessel led to breaking up of the vessel and the formation of an 850-kg, 0.4-m-diameter missile. What fraction of the original water slug energy would have to be imparted to this missile to cause it to break the containment?

BIBLIOGRAPHY

Barsell, A.W. (1981). *A Study of the Risk Due to Accidents in Nuclear Power Plants.* German Risk Study—Main Report, various pages.

Dunster,. J. (1982). "The Assessment of the Risks of Energy." *Atom* (303): 2–6.

Farmer, F.R. (1981). "The Assessment of Risk in Relation to Major Hazards, with Particular Reference to Nuclear Reactors." *Contemp. Phys.* 22 (3): 349–60.

Gittus, J.H. (1982). "International Experience and Status of Fuel Element Performance and Modelling for Water Reactors." Report. ND-R-604(S), U.K. Atomic Energy Authority, UKAEA Risley Nuclear Power Development Establishment, 115 pp.

Green, A.E. and A.J. Bourne (1972). *Reliability Technology.* Wiley Interscience, New York, 636 pp.

Griffiths, R.F. (1978). "Reactor Accidents and the Environment." *Atom,* December, 314–25.

Jones, 0. C. (1981). *Nuclear Reactor Safety Heat Transfer.* Papers presented at the summer school on nuclear reactor safety heat transfer, Dubrovnik, Yugoslavia. August 24–29, 1980. Hemisphere, Washington, D.C., 959 pp.

Kelly, G. N. (1981). "The Radiological Consequences of Notional Accidental Releases from Fast Breeder Reactors." *Ann. Nucl. Energy,* 8 (7): 307–18.

Merilo, M. (1983). *Thermal-Hydraulics of Nuclear Reactors,* Papers presented at the Second International Topical Meeting on Nuclear Reactor Thermal-Hydraulics, Santa Barbara, Calif., January 11–14, 1983, 1529 pp.

Muller, U., and C. Gunther (1982). "Post Accident Debris Cooling." In *Proceedings of the Fifth Post Accident Heat Removal Information Exchange Meeting,* Karlsruhe, West Germany, July 28–30, 1982. Braun, 364 pp.

7

Cooling during Fuel Removal and Processing

7.1 INTRODUCTION

In Chapters 4–6 we have discussed hypothetical and actual accident conditions in reactors. Now we return to the discussion of the next phase of *normal* operation, namely, the removal of the used fuel from the reactor and its subsequent processing.

In a nuclear reactor, the fissile material is gradually used up and converted to energy and fission products. During the nuclear reaction there are changes in the microstructure of the fuel due to the release of fission products, which either combine with the fuel or are released inside the fuel can. These changes have two effects: (1) a gradual deformation of the fuel and in some cases the can and (2) the release of fission products (such as xenon and iodine), which are themselves strong absorbers of neutrons, leading to a reduction in neutron population and a less efficient nuclear reaction. For these reasons, the fuel element must be removed from the reactor after a period of time and before all the fissile material is used up. Typically this period will be between 3 and 5 years for thermal reactors and 1 year to 18 months for fast reactors. For thermal reactors, 60 to 75% of the original fissile material is used up at the time of fuel removal. For the fast reactor, the utilization is much less, of the order of 25%. The fraction utilized is often referred to as the *burn-up*.

The fuel removed from a nuclear reactor contains three kinds of valuable material:

1. The unused proportion of the fissile material that was originally introduced with the fresh fuel.
2. New fissile material that has been bred as a result of the nuclear reactions, in particular, the reaction between neutrons and ^{238}U to form ^{239}Pu. The plutonium produced can be used as a fissile material in both thermal and fast re-

234

actors. Note that bred material also participates in the fission reaction while the fuel is still in the reactor; in a thermal reactor system, 25% of the heat production might arise from fission of material bred in situ.

3. Much of the original ^{238}U, the nonfissile isotope of uranium, still remains. This material is valuable as a *fertile* material for use, particularly in the blankets of fast reactors, where it is converted to ^{239}Pu.

Of course, these valuable materials are mixed with a range of highly radioactive fission products that form the waste from the nuclear cycle. Basically, there are two choices facing the nuclear power operator:

1. To discharge the fuel and store it safely without making any attempt to separate the useful fissile-fertile materials from the fission products in the foreseeable future. If such storage is regarded as permanent, this approach is colloquially referred to as the *throwaway cycle* in the sense that valuable resources are being disposed of. Such a cycle could be practical only if it was felt that the world's uranium resources were adequate to operate thermal reactors for a sufficiently long period. As discussed in Chapter 1, this would be a highly inefficient use of these resources.

2. To discharge the fuel, store it for a relatively short time (typically 1–5 years) to allow the more active fission products to decay and the decay heat to drop to manageable levels, and then to process the fuel chemically to separate the valuable fissile and fertile materials from the fission products, which can then be stored in a safe form.

If a program of fast reactor operation is envisaged, the second option is mandatory; otherwise, far too much of the fissile material in the cycle will be wasted. Reprocessing the fuel is more expensive in the short term than simply storing it, and the decision about whether to reprocess in the case of thermal reactor fuels is closely related to the overall utilization strategy for nuclear energy in individual countries. Where a program of fast reactors is envisaged, reprocessing of thermal reactor fuel is necessary in order to produce the initial inventory of plutonium for such a program. Typically, it would take about 15 years' worth of spent fuel from a thermal reactor to produce the initial inventory for a fast reactor of similar size.

In this chapter we discuss the removal of spent fuel elements from reactors, their transport to a long-term storage location or a reprocessing plant, and the problems of the reprocessing plant itself. The questions of long-term storage of nuclear waste products will be discussed in Chapter 8. This chapter concentrates on the thermal aspects of these operations, in line with the rest of this book.

7.2   REFUELING

A basic decision in the design of any nuclear reactor is whether to remove and insert fuel while the reactor is operating (*on-load refueling*) or when the reactor is shut down (*off-load refueling*).

The choice between on-load and off-load refueling is dictated primarily by economic factors. On-load refueling is much more complex, and the cost of the equipment is high. On the other hand, total shutdown of the reactor system for a significant period to allow off-load refueling leads to a loss of electrical power output, and this in itself is very expensive. In general, reactors that have a large throughput of fuel, such as natural uranium reactors (Magnox and CANDU) operate with on-load refueling, whereas those with a lower throughput of enriched fuel (e.g., PWR and BWR) tend to use off-load refueling. The advanced gas-cooled reactor (AGR) is intermediate between these two cases; it has the capability of on-load refueling, though this is only just being introduced into routine operations (Jenkins et al., 1995). For fast reactors, off-load refueling is necessary because of the very large changes in reactivity that would occur with any fuel movements during operation.

7.2.1   Refueling of Gas-Cooled Reactors

Early gas-cooled reactors (the air-cooled piles such as those at Windscale in the United Kingdom) had horizontal channels, and the fuel elements were simply pushed in at one end and spent fuel was removed at the other. With the introduction of Magnox reactors, which had vertical channels and used a pressurized carbon dioxide coolant, this simple system was no longer adequate. The refueling arrangements used for a Magnox reactor are illustrated in Figure 7.1. An array of vertical pipe comes from the top of the reactor vessel as illustrated (these are called *standpipes*). The refueling machine may be connected to any of the standpipes. This machine is shown in Figure 7.1 and is basically a pressure vessel that can be moved across the top face of the reactor. It is provided with adequate radiation shielding and therefore tends to be heavy. When the refueling machine is connected to one of the standpipes, a plug is removed from the top of the standpipe, allowing the high-pressure carbon dioxide coolant to enter the refueling machine vessel; thus, the vessel becomes an extension of the primary circuit of the reactor. Each standpipe serves a group of fuel channels. The fuel elements are lifted out of the channels using a grab, which is aligned above the particular channel using a special mechanism called a *pantograph* (Figure 7.1).

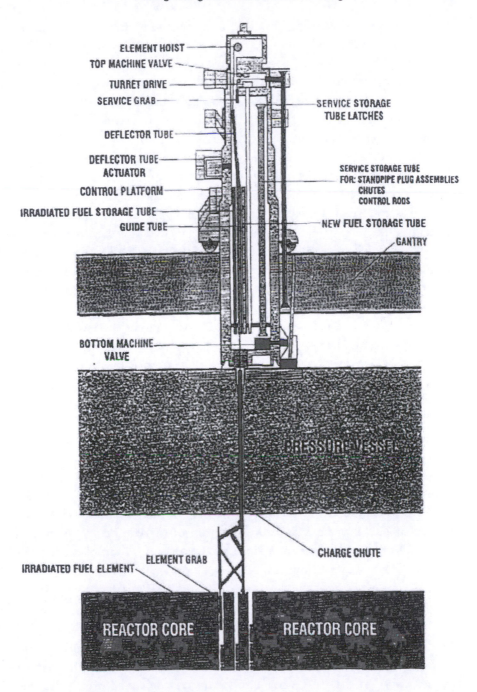

Figure 7.1: Magnox refueling machine.

Alternatively, in some cases, an aligning chute is used. In a typical refueling operation, all of the individual fuel elements in a fuel channel are removed and stored temporarily in magazines in the refueling machine vessel. New fuel elements, already present in the vessel, are then inserted using the same mechanism. In the Magnox reactor, no special cooling is provided for the spent fuel elements within the refueling machine since natural convection of gas around the elements keeps them cool enough.

In the AGR the arrangement is similar to that used in the Magnox reactor, except that there is a standpipe for every fuel channel as illustrated in Figure 7.2. Thus, the complete fuel from the channel can be drawn up into the refueling machine as a single entity, and the complex pantograph or chute mechanism is avoided. In the AGR the fuel rating is much higher, and thus the decay heat release rate is such that natural convection cooling of the spent fuel within the refueling machine may be insufficient. During the refueling operation, therefore, carbon dioxide from the reactor circuit is passed through the refueling machine and over the spent fuel. The fuel damage incident at Hinkley Point B, described in Section 5.4.4, led to increased attention to cooling during the refueling operation and to the installation of backup emergency cooling systems.

In both Magnox and AGR reactors, the refueling machine containing the spent fuel is trundled over to a discharge point where the magazines are emptied into an irradiated fuel buffer store that is gas-cooled. Subsequently, they may be transferred (also using the refueling machine) to a more permanent storage at the reactor (normally a deep pool of water) before being finally transported from the site. The sequence for an AGR is illustrated in Figure 7.3.

A gas-cooled reactor that we have not previously mentioned and that has a novel method of on-load refueling is the *pebble-bed reactor* developed in the Federal Republic of Germany. In this reactor the fuel is incorporated into graphite spheres that are charged into the top of the reactor, the spent fuel being discharged at the bottom. A small prototype of this form of reactor was operated for a considerable time.

7.2.2   Refueling of CANDU Reactors

The diagram of the CANDU in Figure 3.6 shows the positioning of the two refueling machines at either end of one of the horizontal channels. Each machine is a pressure vessel that can be connected to the ends of the horizontal channel, becoming pressurized to system pressure when a plug at the end of the chan-

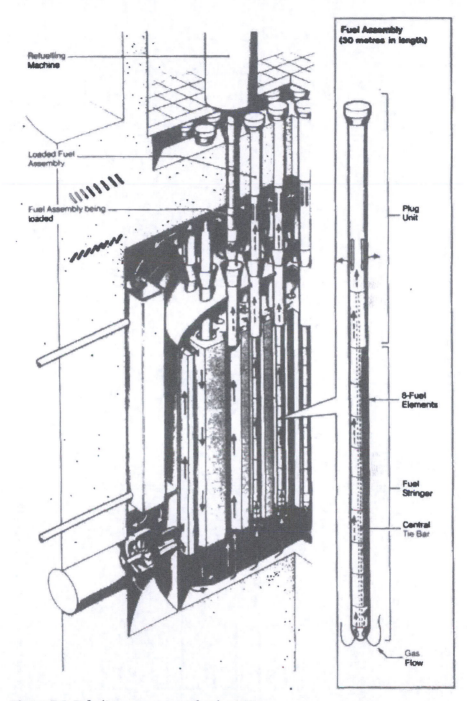

Figure 7.2: Refueling arrangement for the AGR.

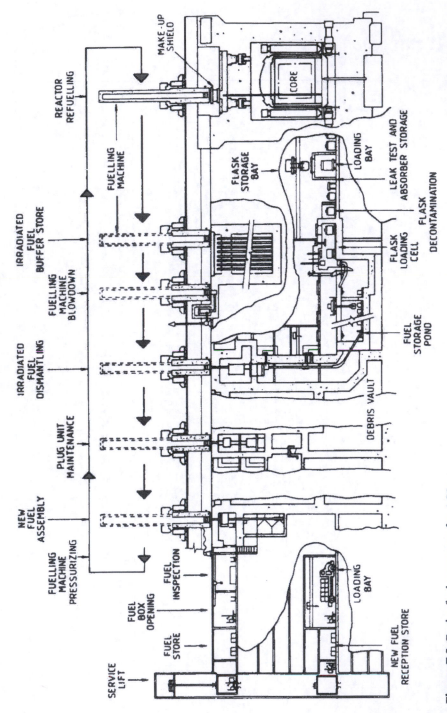

Figure 7.3: Fuel-refuel sequence for AGRs.

nel is removed. Each refueling machine contains a magazine that can hold either spent fuel (at the discharge end) or fresh fuel (at the inlet end). A ram is used to push the fuel bundles through the channel. The success of these refueling machines has contributed significantly to the very high *load factor* (proportion of time for which the reactor is at power) achieved in the CANDU reactors as a result of on-load refueling. For a typical 600-MW(e) reactor, approximately 70 fuel bundles are changed each week. The fuel in the machines is cooled by means of a flow of heavy water taken from the main reactor coolant circuit and passed through the machines back into the fuel channels.

7.2.3 Refueling of Light-Water Reactors

In the case of the PWR and BWR, the refueling is off-load. It takes place approximately once a year over a period of 4–6 weeks. Other maintenance work on the plant is scheduled to be done at the same time, which means that high load factors are still achievable with these reactors.

To carry out refueling in a PWR or BWR, the system is partially drained to bring the liquid level to below the level of the flange that connects the main part of the vessel to the top part (referred to as the *top head*). All the control rods are fully inserted into the core and unlatched from their mechanisms (which pass through the head). The bolts attaching the top head to the vessel are then loosened, the cavity in which the reactor sits is flooded with water, and the head is removed. The upper structures in the reactor vessel are removed to expose the fuel, and handling operations are carried out under a significant depth of water in the reactor cavity (typically 5–10 m). This water is also circulated through a heat exchanger to provide liquid cooling for decay heat removal. Approximately one-third of the total number of fuel elements are removed in any one operation, namely, about 50–60 elements out of the total inventory of 200.

As shown in Figure 7.4, in the case of the PWR, the fuel is passed into a *transfer canal*, in which it is transferred horizontally out of the reactor building and into a water-filled fuel storage pond.

The refueling route for the BWR is similar to that illustrated in Figure 7.4, but with the additional complication that it is necessary to remove all of the devices above the core used to separate the steam from the steam-water mixture leaving the core (see Figure 4.27 for an illustration of the reactor structure). The BWR fuel elements are somewhat smaller than those in the PWR, and therefore a correspondingly larger number of fuel movements must be made.

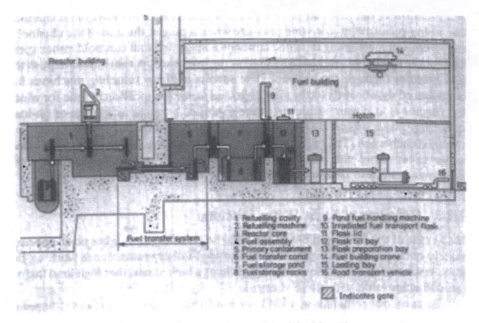

Figure 7.4: Sizewell B power station PWR irradiated fuel handling route.

Once every 3 years it is common practice to remove all the fuel and the lower core structures and to carry out a thorough inspection of the pressure vessel from the inside surfaces. This provides a guarantee of the integrity of this vessel, which is essential to the safety of the system. The internal structures and fuel are then recharged into the vessel and the reactor restarted. Typically, this triennial inspection process might take up to 3 months.

7.2.4   Refueling of Liquid Metal–Cooled Fast Breeder Reactors

Figure 7.5 shows the refueling route for a large pool-type fast breeder reactor. The objective in this refueling process is to keep the used fuel permanently under sodium, which acts as a heat sink for the decay heat. The fuel is extracted by a grab attached to a rotating plate above the reactor. It is extracted from the core and, still under sodium, is passed into an intermediate buffer store, from which it is transferred through a sloping transfer line (also sodium-filled) to a sodium-cooled spent fuel store, where it is kept for 100–200 days before being transferred to the reprocessing plant.

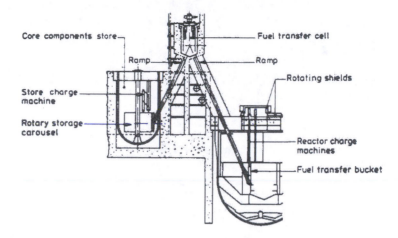

Core components store—

Fuel transfer cell

Ramp

Ramp

Rotating shields

Store charge machine

Rotary storage carousel

Reactor charge machines

Fuel transfer bucket

Figure 7.5: Refueling route for large pool-type fast reactors.

7.3 SPENT FUEL STORAGE AND TRANSPORT

The complete cycle for nuclear reactor fuel (the *fuel cycle*) is illustrated in Figure 7.6. As will be seen, storage and transport of irradiated fuel play an important role in this cycle.

As we saw earlier, nuclear reactor fuel continues to emit heat even after the fission reaction ceases, due to fission product decay heating. Figure 7.7 shows the heat release rate as a function of time for spent fuel from the various types of reactors. Clearly, the more highly rated the reactor (e.g., the fast reactor), the higher the heat release rate and the longer it takes for it to decay to a low value.

Figure 7.7 shows that the fission product heat release is most intense immediately after discharge. This is why it is common practice to store the fuel in a *cooling pond* for a period of time to allow both the radioactivity and the heat release to decay before removing the fuel from the immediate environment of the reactor. It is usual to store the fuel at the reactor site in a pool of water (though not, obviously, for the fast reactor fuel), although some air-cooled and gas-cooled (carbon dioxide) stores have been designed and operated. Water pools are well suited for fuel designed for water-cooled reactors, but they present a difficulty for the storage of fuel whose cladding has been designed for satisfactory performance in a gas environment. For example, the immersion of Magnox fuel for long periods in water ponds allows a slow chemical reaction to occur between the magnesium alloy cladding and the water, and this leads to the generation of

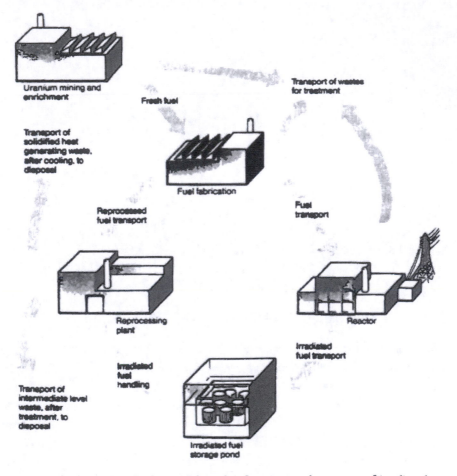

Figure 7.6: The fuel cycle showing the role of storage and transport of irradiated fuel.

hydrogen and the formation of a potentially troublesome silt of radioactive magnesium hydroxide. If the can is severely corroded, fission products may escape from the fuel into the pond, giving environmental control difficulties. However, with good management of the ponds (including special encapsulation of fuel that is known to be damaged), these effects can be minimized.

As with all other aspects of nuclear power, consideration must be given to the safety of the operation of spent fuel storage ponds. This can be illustrated by considering PWR fuel assemblies, which are unloaded from the reactor and may be stored in water ponds for many years. The decay heat levels of PWR fuel assemblies are such that if the water is completely drained from the pool,

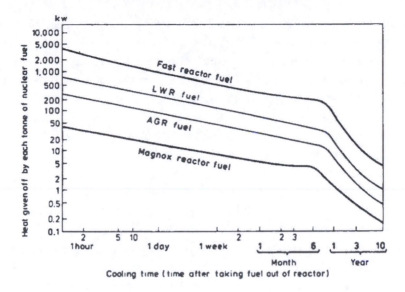

Figure 7.7: Heat release from spent nuclear fuel.

the fuel that has been out of the reactor for fewer than 150 days will melt. Loss of water from the pool could occur if the pool developed a leak or if the pool cooling system were turned off, leading to water evaporation. Both of these events are extremely unlikely. However, the defense-in-depth strategy is continued at the storage stage by either placing the store within the reactor containment (as is done in the German PWR designs) or by providing it with its own containment, including ventilation and filtration systems (the U.S. approach). The pond water is cooled by passing it through heat exchangers, and failure of this cooling system is perhaps the most likely failure mechanism for the ponds. However, it is unlikely that the operators would not notice a gradual fall in the water level in the ponds over a period of about 2 weeks, which would be required to uncover the fuel by evaporation due to the heat input from the fuel itself. Thus, loss-of-coolant accidents in fuel ponds are considered minor contributors to the overall risks of nuclear power.

In designing storage ponds for nuclear reactor spend fuel, consideration must be given to the problem of *criticality*, that is, the possibility that the pond itself would act as a nuclear reactor. With natural uranium fuel (Magnox and CANDU) there is no criticality problem in storing the fuel under water since the natural uranium–light water system does not become critical. For PWR, BWR, and AGR spent fuel, it is hypothetically possible to have a nuclear reaction with the fuel placed in a water pool. Thus, the pools must be designed with sufficient distance between the fuel elements to guarantee that no reaction occurs. The distance between the fuel elements in the store can be reduced if neutron-absorbing material is interspersed between the individual subassembly channels, allowing a much higher packing density in a pool.

From a typical 1000-MW(e) PWR, about 25 tons of fuel are discharged every year, contained in about 60 fuel assemblies. About 8000 tons of spent fuel are removed from power reactors each year in OECD countries and some 150,000 tons of spent fuel are currently in storage ponds. With this rate of discharge, it is obvious that after a number of years the storage facilities at reactor sites will become full and fuel will have to be transported either to an alternative storage site or to a reprocessing plant.

Spent nuclear fuel is transported by placing one or more fuel assemblies in a *transport flask*, in which a large number of assemblies are transferred in a water-filled basket. A typical transport flask (or *cask* in U.S. terminology) for water reactor fuel is illustrated in Figure 7.8. Figure 7.9 illustrates the spent fuel flask used for Magnox fuel; the fuel is contained in a water-filled box (*skip*) sur-

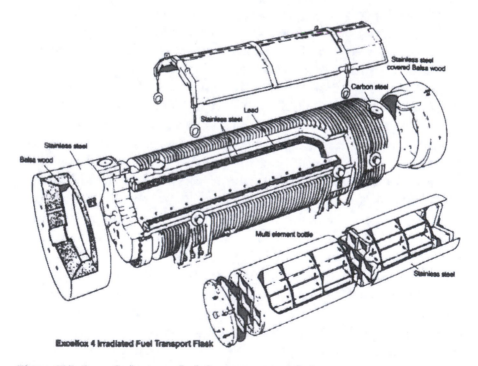

Figure 7.8: Spent fuel storage flask for water reactor fuel.

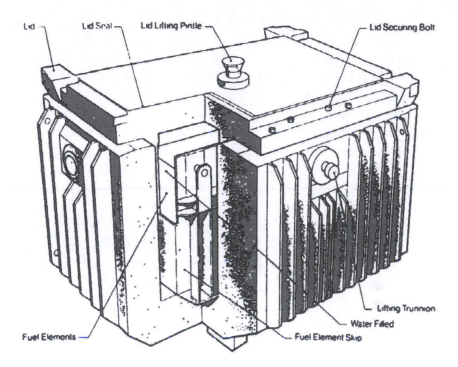

Figure 7.9: Spent fuel flask used for the transport of spent Magnox fuel.

rounded by the flask shielding. The fuel is placed in a steel basket inside the flask, which is then sealed with a cover as shown. The flask wall has a series of layers as illustrated in Figure 7.8 with a 12–14-in.-thick outer steel layer and inner layers of depleted uranium and/or lead to absorb the gamma radiation and of water to act as a neutron shield. A flask for road transport might weigh about 20 tons and contain one or two elements, whereas a flask for rail transport might be much bigger, weighing up to 100 tons and able to carry 10–20 fuel assemblies.

During transport, heat must be dissipated from the outside surface of the cask. Typical heat dissipation rates would be about 10 kW for a road transport cask and 50–100 kW for the large rail transport cask. There are two main steps in this heat transfer process. First, heat is transferred from the fuel to a fluid within the flask (usually water), which circulates by natural convection around the fuel. The heat is then taken from the water into the flask wall and out to the atmosphere. The flasks normally have steel fins on the outside to assist the heat dissipation to the air.

A variety of accidents involving transport flasks can be postulated. First, they may be accidentally dropped during transfer from the storage pool to the vehicle. To withstand such an impact, the flask must be designed to survive a drop of 30 ft onto an unyielding (e.g., concrete) surface without any impairment of its integrity and also survive a 40-in. drop onto a 6-in. spike. Second, the flask may become involved in a fire, and prototypes of a given design of flask are subjected to tests in which they are placed in a fire at 1000°C for a period of 30 min. Survival of these stringent tests is a necessary condition for licensing. Apart from these standard tests, demonstrations have been carried out by CEGB in the United Kingdom and at Sandia Laboratories in Albuquerque, New Mexico, in which simulated accidents have been staged. For instance, the effect of a low-loader truck with a transport flask on it, stationary on a railway crossing, being hit by a locomotive traveling at 100 mph. has been examined. The fact that the flask survived such dramatic impacts unscathed (although the locomotive did not!) has inspired great confidence in the safety of transporting spent nuclear fuel in this way.

7.4   REPROCESSING PLANT

If it has been decided to reprocess spent fuel with the objective of recovering valuable uranium and plutonium, the fuel must first be transported to a reprocessing plant using the flasks described in the previous section. The stages that the fuel then goes through in the separation process are illustrated schematically in Figure 7.10. First, the flask is taken off the vehicle, the spent fuel is removed under water, and the flask is decontaminated and returned to the power station for further use. The fuel is loaded into a storage rack under water until it is ready to be fed into the reprocessing plant.

In a modern reprocessing plant like THORP (Thermal Oxide Reprocessing Plant) operated by British Nuclear Fuels at Sellafield, the actual separation process is undertaken after at least 5 years' storage of the spent fuel in the ponds. The fuel element is first stripped of as much of its extraneous metal structure (grids, support plates, etc.) as possible. These remnants are stored separately and treated as intermediate-level waste (see Chapter 8). The fuel pins themselves are sheared into small lengths between 1 and 4 in.; these sheared fuel pieces fall down a chute into a perforated basket (see Figure 7.10). This basket is then transferred to the dissolver. The shear needs to be of modular construction to allow replacement of the blade and for maintenance.

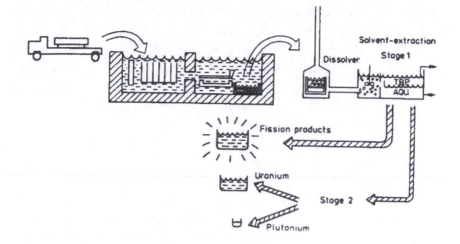

Figure 7.10: Schematic diagram of reprocessing plant.

In the dissolver the fuel is dissolved in hot (90°C) 7 M nitric acid. Dissolution of the fuel takes place quickly and can be controlled by the rate of shearing. The cladding pieces, or "hulls," are withdrawn in the basket and again sent for disposal as intermediate-level radioactive waste. Various types of dissolver, both batch and continuous, have been developed. As the fuel dissolves, fission gases are released: the inert gases krypton and xenon and other volatiles such as iodine and carbon dioxide as well as oxides of nitrogen and steam. The dissolver off-gas systems must be able to cope with this mixture. The system recovers as much of the nitrogen oxides as possible as nitric acid.

The fuel solution itself still contains some undissolved particulates, both from the cladding and from fission products. The solution is therefore clarified using a centrifuge. The clarified nitric acid solution containing the fission products, the uranium, and the plutonium is next passed through the chemical separation plant. This involves a solvent extraction system.

Solvent extraction is a process that allows separation of dissolved materials. Suppose we have two liquids that do not mix, such as oil and water. If we have a solution of two substances, A and B, in one of the liquids, and component B is soluble in the other liquid but component A is not, then we may solvent-extract component B from the original mixed solution of A and B by essentially shaking up ("contacting") the solution with an immiscible liquid in which only B is soluble. By then removing component B from the resultant solution, we have achieved a separation of A and B. Various types of equipment are used in chemical engineering for this process, and it is beyond the scope of this book to go into them in detail. Probably the most commonly used devices in reprocessing plants use mechanical stirrers to mix the two liquids, followed by settling tanks that allow their separation, with each of the liquids containing the respective components. These are called *mixer settlers.* Alternatively, vertical pipes containing perforated metal plates may be used, with one fluid flowing up the pipe and the other flowing down it. To promote mixing of the fluids, such columns are subjected to pulses, and they are often referred to as *pulsed columns.* A typical pulsed column is shown in Figure 7.11. The first objective of solvent extraction in the reprocessing plant is the separation of the valuable uranium-plutonium mixture from the nitric acid solution, which also contains the fission products. This is done by contacting the nitric acid fuel solution with an organic solvent, typically tributyl phosphate (TBP) diluted with odorless kerosene (OK). In a typical extraction plant, all but about 0.1% of the uranium and plutonium in the fuel solution is removed into the TBP phase.

Separation of the uranium from the plutonium is also achieved by solvent extraction. The first step is to redissolve the mixture in a clean acid stream and then add a substance to the stream to change the condition of the plutonium and render it insoluble in TBP. Thus, when the new acid stream is contacted again with the TBP, the plutonium remains in the acid stream while the uranium passes into the TBP. The success of the extraction process is largely dependent on the efficiency of the transfer from the aqueous phase and vice versa. In general, the uranium-plutonium will dissolve preferentially in the TBP when the aqueous phase has a high nitric acid content and will dissolve preferentially in the aqueous phase when it has a low nitric acid content. Thus, the final stage of the extraction is to take the uranium from the TBP stream by contacting the stream again with an aqueous phase having a low concentration of nitric acid.

The output of the separation stages in the reprocessing plant consists of streams of uranium, plutonium, and fission products dissolved in nitric acid. Each of these streams may be concentrated by evaporation and subsequently purified, if necessary, by additional solvent extraction stages. The uranium and plutonium are precipitated as uranium and plutonium nitrates, which are then heated to convert them into oxides, which may then be reused in the preparation of nuclear fuel. The fission product stream is usually concentrated by evaporation and passed to storage tanks for long-term storage and ultimate conversion into a solid form; we shall discuss this process in Chapter 8.

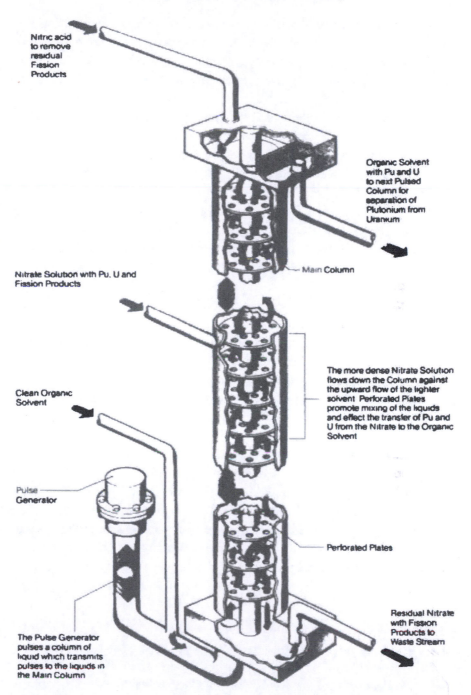

Nitric acid to remove residual Fission Products

Organic Solvent with Pu and U to next Pulsed Column for separation of Plutonium from Uranium

Nitrate Solution with Pu, U and Fission Products

Main Column

The more dense Nitrate Solution flows down the Column against the upward flow of the lighter solvent Perforated Plates promote mixing of the liquids and effect the transfer of Pu and U from the Nitrate to the Organic Solvent

Clean Organic Solvent

Pulse Generator

Perforated Plates

Residual Nitrate with Fission Products to Waste Stream

The Pulse Generator pulses a column of liquid which transmits pulses to the liquids in the Main Column

Figure 7.11: Typical pulsed column used for solvent extraction of fission products from spent fuel.

Once the uranium and plutonium have been extracted, the decay heat generation is almost totally associated with the fission product stream in the reprocessing plant. Any heat transferred to the solvent phase, together with the intense radiation, tends to degrade the solvent and cause difficulties in the operation of the plant.

The thermal and radiation problems in reprocessing plants are obviously fewer the longer the fuel has been stored in the cooling ponds prior to reprocessing. It is for this reason that for thermal reactor systems the storage period is 5 years or more. However, this is not possible for fast reactors, where the economics of the fuel cycle dictates a fast turnaround in reprocessing. Much more fissile material is contained in fast reactor fuel than in thermal reactor fuel, and failure to utilize this valuable capital resource results in a considerable economic penalty. Furthermore, the rate at which fast reactors can be built is limited because of the very much larger total inventory of valuable fissile material associated with each reactor.

Therefore, fast reactors present greater difficulties for reprocessing than do thermal reactors. They already have a higher specific heat generation rate, as seen in Figure 7.7, and their spent fuel must be reprocessed on a much shorter time scale, typically 6–9 months after removal from the reactor. The very high concentration of fissile materials in the streams presents a further difficulty. In the design of a reprocessing plant for *both* thermal and fast reactor fuel, one must take into account the possibility of developing a nuclear reaction (criticality) within the plant. This can be prevented in many cases by designing the plant so that the geometry of the pipes containing the solutions of fissile material is so unfavorable to the nuclear reaction that the plant can be regarded as "ever-safe." This is particularly important in the reprocessing of fast reactor fuels where the concentrations of fissile material are high and the throughputs are small. Such plants are successful when proper attention is given to the design details; an example is the U.K. Atomic Energy Authority's fast reactor fuel reprocessing plant at Dounreay in Scotland, which is illustrated schematically in Figure 7.12.

REFERENCES

Jenkins, G.E.C., Lee, M.D., and N. Wall (1995). "Improved Refueling of Advanced Gas-Cooled Reactors." *Nuclear Europe Worldscan* 15 (March–April): 44.

Nuclear Fuel Reprocessing Technology (1985), published by British Nuclear Fuels plc, Information Services, Risley, Warrington, U.K.

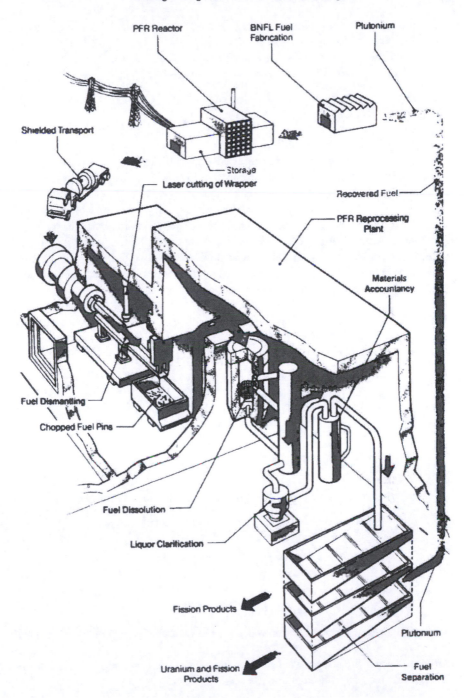

PFR Reactor

BNFL Fuel Fabrication

Plutonium

Shielded Transport

Storage

Laser cutting of Wrapper

Recovered Fuel

PFR Reprocessing Plant

Materials Accountancy

Fuel Dismantling

Chopped Fuel Pins

Fuel Dissolution

Liquor Clarification

Fission Products

Plutonium

Fuel Separation

Uranium and Fission Products

Figure 7.12: Reprocessing plant for the U.K. prototype fast reactor at Dounreay, Scotland.

Fuel Reprocessing Services (1986), published by British Nuclear Fuels plc, Information Services, Risley, Warrington, U.K.

Fuel Handling and Site Ion Exchange Effluent Plant (1985), published by British Nuclear Fuels plc, Information Services, Risley, Warrington, U.K.

EXAMPLES AND PROBLEMS

1 Loss of water from cooling pond

Example: The fuel elements arising from an LWR refueling are placed in a cooling pond on removal from the reactor. Twenty-five tons of fuel are placed in the pond, which is 10 m wide, 20 m long, and has a water depth of 10 m. After 1 month of storage, there is a failure of normal water supply to the pond. How long would it take for a 0.5-m drop in water level to occur due to evaporation? Assume that the water temperature at the time of the supply failure is 25°C and that the fuel element volume is negligible compared with the pool water volume. Assume that the water has a specific heat capacity of 4.18 kJ/kg K, a density of 1000 kg/m^3, and a latent heat of evaporation of 2.25 mJ/kg. Assume that there are no heat losses from the pool.

Solution: The heat release rate per ton of fuel after 1 month is given by Figure 7.7 and is 70 kW. The total heat release rate from the fuel is thus $25 \times 70 = 1750$ kW. With no water throughflow in the pool, the water will first rise to the boiling point. The amount of heat required to bring the water to its boiling point (100°C) is given by

$$\text{Volume of pond density} \times \text{specific heat} \times (100 - 25)$$
$$= (10 \times 20 \times 10) \times 1000 \times 4.18 \times 10^3 \times 75$$
$$= 6.27 \times 10^{11} \text{J}$$

The time t_b required to reach boiling point is thus

$$t_b = 6.27 \times 10^{11} \div 1750 \times 10^3 = 3.58 \times 10^5 \text{s}$$
$$= 99.5 \text{h}$$

The time t_e for the depth of water to fall by 0.5 m due to evaporation is given by

$$t_e = \frac{\text{volume of water evaporated} \times \text{density} \times \text{latent heat}}{\text{heat release rate from fuel}}$$
$$= \frac{(10 \times 20 \times 0.5) \times 1000 \times 2.25 \times 10^6}{1750 \times 10^3}$$
$$= 1.29 \times 10^5 \text{s}$$
$$= 35.7 \text{h}$$

Thus, the total time required for the water level to fall by 0.5 m is $36 + 99 = 135$ h (5.6 days).

Problem: Repeat the above calculations assuming that the water supply failure occurs 1 week after the fuel is inserted into the pond.

2 Heat losses from a fuel flask

Example: A cylindrical fuel flask is to be designed to transport complete fuel assemblies from a thermal reactor to a reprocessing plant after their removal from the cooling pond. The fuel assembly contains 300 kg of spent fuel, which at thc time of its removal from the pond is releasing 3W/kg of heat. The fuel flask is 0.6 m in outside diameter and 3 m long. If the outside temperature of the flask is to be maintained at less than 10°C above ambient temperature, will the rate of heat transfer to the atmosphere be sufficient without finning the outside of the flask? If the cooling is insufficient, calculate the number of vertical 3-m-long, 10-cm-high fins that would have to be attached to the surface to maintain the desired cooling. Assume a heat transfer coefficient α between the flask surface (and this fin surface) and the atmosphere of 10 W/m² K.

Solution: The total heat generation rate is $300 \times 3 = 900$ W, and the heat flux $\dot{q}$ from the surface is given by

$$\dot{q} = \frac{900}{\pi \times 0.6 \times 3} = 159 \text{ W/m}^2$$

The temperature difference between the flask surface and the atmosphere is thus $\dot{q}/\alpha$ = 15.9K, which is greater than can be accepted. For a temperature difference of 10K (10°C), the flask surface can lose $10 \times 10 \times \pi \times 0.6 \times 3 = 565$ W. The additional surface area required is thus $(900 - 565) \div 10 \div 10 = 3.35$m². The area per fin is given by $3 \times 0.1 \times 2 = 0.6$ m². Thus, 6 vertical fins placed around the circumference of the flask would be sufficient to maintain the temperature of the outer surface at less than 10°C above ambient.

Problem: The fuel flask described in the example is being considered for transport of fuel elements at an earlier stage of their cooling, where the decay heat release rate is 10W/kg. If the temperature on the outside of the flask was allowed to be 15°C above ambient, would additional fins be required and, if so. how many?

3 Heat generation in reprocessing plant streams

Example: After 3 years of storage, a spent fuel emitting 2W/kg of heat due to fission product decay is reprocessed, the fission products being extracted into an aqueous nitric acid raffinate stream. Internal heating of the stream occurs due to fission product decay at a rate of 400 W/m³. The stream is passed from the reprocessing plant to a storage tank along a 1-cm-diameter, 10-m-long pipe at a rate of 30 liters/h. Assuming that the inlet temperature to this pipe is 25°C and that heat losses to the atmosphere from the pipe are negligible, calculate the temperature of the stream entering the storage vessel. The specific heat capacity of the stream should be taken as 4 kJ/kg K and its density as 1200 kg/m³.

Solution: The velocity v of the raffinate stream throught the pipe is given by

$$v = \frac{\text{volume flow (m}^3\text{/s)}}{\text{cross-sectional area of pipe}}$$
$$= \frac{30 \div 3600 \div 1000}{\pi \times 0.01^2 / 4}$$
$$= 0.1061 \text{ m/s}$$

Assuming (incorrectly!) a uniform velocity in the tube, the time taken for an element of fluid to travel through the 10-m length of tube is 10/0.1061 = 94.2 s. During this time, 94.2 × 400 J/m^3 = 377 × 10^4 J/m^3 of heat is added to the stream. This will give a temperature rise ΔT the stream that is calculated from

$$\Delta T = \frac{\text{heat added per cubic meter}}{\text{density} \times \text{specific heat}}$$

$$= \frac{3.77 \times 10^4 \text{ J/m}^3}{1200 \times 4 \times 10^3}$$

$$= 0.0078°C$$

Thus, the temperature of this stream entering the storage vessel is 25.0078°C, only slightly higher than the inlet temperature.

Problem: Suppose that a flow blockage occurs in the transfer line described in the example and the flow stops, but the fluid remains in the line. The reprocessing plant is shut down. Assuming, again, that heat losses from the line are negligible, calculate the time taken for the fluid to reach its boiling point of 95°C due to fission product decay heating.

BIBLIOGRAPHY

Allardice, R.H., and D.W. Harris (1981). "Fast Reactor Fuel Reprocessing." *Nucl. Energy* 20 (1): 63–69.

Avery, D.G., and D.O. Pickman (1976). "Development and Production of Nuclear Fuel in the UK." Report TRG 2801(S), UKAEA Risley Nuclear Power Development Establishment, 17 pp.

Chayes, A., and W.B. Lewis (1977). *International Arrangements for Nuclear Fuel Reprocessing.* Ballinger, Cambridge, Mass., 251 pp.

Chicken, I.C., et al. (1981). "The Environmental Impact of Transporting Radiative Materials." In *Environmental Impact of Nuclear Power, Proceedings of a Conference,* British Nuclear Energy Society, London, 294 pp.

Collier, J.G. (1981). "The Nuclear Fuel Cycle and Proliferation." In *Environmental Impact of Nuclear Power, Proceedings of a Conference,* April 1–2, 1981, British Nuclear Energy Society, London, 294 pp.

Marshall, W., ed. (1983). *Nuclear Power Technology,* vol. 2, *Fuel Cycle.* Clarendon, Oxford, 3 vols.

Mathews, R.R. (1980). *Transport of Irradiated Nuclear Fuel.* Central Electricity Generating Board, London, 13 pp.

Messenger, W. de L.M. (1979). "The Safe Transport of Radioactive Materials." Report ND-R-259(R),UKAEA Risley Nuclear Power Development Establishment, 17 pp.

Parker, R.J. (1978). *The Windscale Inquiry – Report,* HMSO, London, 3 vols.

Shortis, L.P., et al. (1964). "Reprocessing of Irradiated Fuel in the United Kingdom." Paper 9, Session 4 of the Anglo-Japanese Nuclear Power Symposium, Tokyo, March 1963.

8

Cooling and Disposing of the Waste

8.1 INTRODUCTION

All forms of energy production result in the formation of waste, the safe management of which is essential for the protection of the public and the environment. These wastes may be produced at various stages of the fuel cycle: extraction, refining, and utilization. In the case of the use of fossil fuels the main waste products from combustion are carbon dioxide and the "acid rain" gases: sulphur dioxide and nitrous oxides. Even in the case of "clean" renewable energy sources, waste products associated, for example, with the production of photovoltaic materials or from geothermal systems need to be taken into account in any environmental balance sheet.

Nuclear energy is no exception and waste products are formed at each stage of the fuel cycle. Some of these waste products are radioactive, and it is necessary to handle, store, and dispose of these materials with extreme care. The long-term disposal of radioactive species is an integral part of the design and operation of the nuclear fuel cycle.

8.2 CLASSIFICATION OF WASTE PRODUCTS

Radioactive wastes can arise in gaseous, liquid, or solid forms. In general, at some stage of the management process the radioactivity in the gaseous and liquid forms is converted into a solid form. Most attention is therefore directed at the disposal of solid waste. Critics of nuclear power sometimes refer to radioactive waste disposal and decommissioning as the Achilles' heel of this energy source. In fact, safe, sound, and economic technical solutions have been established for these activities.

Essentially, waste products from nuclear power may arise as follows:

Uranium Mining. The spoil from uranium mining is mildly radioactive and may need stabilization and monitoring.

Fuel Fabrication Plant. The enrichment and fabrication plants for uranium-based fuel present no particular problems in terms of radiation hazard. However, the fabrication of plutonium-based fuel produces low-activity plutonium-bearing residues of wastes arising from the fabrication process.

Spent Nuclear Fuel. As we have seen in Chapter 7, spent nuclear fuel includes the highly radioactive fuel matrix together with the fuel can and supporting grids. The matrix itself contains the highly radioactive fission products, the remaining part of the original fissile and fertile materials, and the material bred in the reaction (see Section 7.1). Even if the fissile and fertile materials are recycled, the highly radioactive fission products remain and are the most important wastes arising from nuclear power. Their disposal will be the main focus of this chapter.

Reprocessing Plant. In addition to the recycled product streams (uranium and plutonium) and the fission product stream, reprocessing produces a number of other waste streams. These include aqueous and organic streams containing medium levels of radioactivity. Another waste product is the residual cladding and support materials from the fuel elements, often referred to as *hulls*. These are reduced by compaction and then stored in a matrix of concrete or bitumen prior to final disposal. Reprocessing plants also generate waste with low levels of active contamination, including rubber gloves, tissues, and plastic containers. Some of these materials are contaminated by plutonium. The disposal of low-level waste materials will be discussed in Section 8.7.

Nuclear Reactors. In addition to the spent fuel, certain other radioactive products come from the reactors themselves. These include gaseous wastes (such as xenon and krypton) that may escape from defective fuel within the reactor, liquid wastes such as tritium oxide (the form of water produced from the tritium isotope of hydrogen), solid wastes such as the resins from the water treatment plants that are used to clean up any small amounts of fission and corrosion products that may enter the primary system, and the filters from the cleanup system in a gas-cooled reactor. Finally, when the reactor comes to the end of its useful life, it must be *decommissioned*: the structural materials will have become slightly radioactive during the operation of the reactor and a careful program of work is needed to return the site safely to normal use. We shall discuss this problem in Section 8.7.

8.3   FISSION PRODUCTS AND THEIR BIOLOGICAL SIGNIFICANCE

In Section 1.4 we described a typical fission reaction, producing atoms of barium-141 and krypton-92 by the fission of a uranium-235 atom. In practice, fission products range in atomic mass from about 80 to 160. For each kilogram of fissile material converted, a certain percentage is converted to one pair of fission products, a certain percentage to another, and so on. The percentages of the fission products formed may be plotted as a function of atomic number (Figure 8.1). Typically, there are about 40 possible fission reactions producing about 80 different species of fission product. The half-lives of these species vary from a fraction of a second to 30 years or more. The short half-life materials are not important since they decay rapidly inside the reactor and during the storage period after removal from the core.

In discussing the significance of radioactive fission products in the environment, it is usual to focus attention on those that are likely to be the most troublesome—in particular, the isotopes that if released would be absorbed and concentrated in specific rgans of the body. For example, various radioactive isotopes of iodine that are formed in the fission reaction, or are subsequently formed by decay of other fission products, can concentrate in the thyroid gland. The iodine isotopes of main interest are I^{131} (half-life, 8 days), I^{132} (half-life, 2.3 h), and I^{129} (half-life, 20 million years). In general, the longer the half-life, the less intense the radiation. In an accidental release it might be expected that iodine would deposit on grassland, be eaten by cows, appear in the milk, and be taken up by people drinking milk, especially children. For this reason, the behavior of iodine has received detailed attention in nuclear safety studies, and there are plans that in the highly improbable event of a serious release, milk from affected areas will be collected and disposed of.

A useful concept in considering the hazards of radioactive fission products is that of *biological half-life*. This is the time needed for any particular radioactive element, taken into the body, to be reduced to half its level of natural excretion

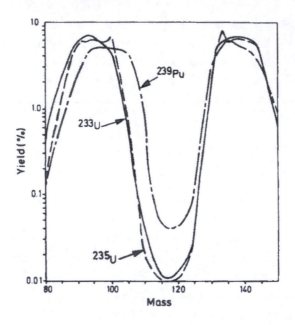

Figure 8.1: Mass-yield curves for thermal-neutron fission of U^{233}, U^{235}, and Pu^{239}.

processes. The significance of this concept can be appreciated by comparing two of the most important fission products, caesium-137 and strontium-90. These isotopes have radioactive half-lives of approximately 30 years. However, the biological half-lives are very different, around 70 days for caesium and 50 years for strontium. The long biological half-life of strontium is due to the fact that it accumulates in the bone structure. Thus, strontium is considered a more serious hazard than caesium.

A material of great interest in radiological protection is plutonium-239, which also has a long biological half-life (200 years in the bone structure and 500 days in the lung). Since the radioactive half-life of plutonium is about 25,000 years, the effective half-life in the body is dominated by the biological half-life.

Another important radioisotope, tritium, is emitted in small quantities from water reactors and reprocessing plants. It is formed by the process of *ternary fission*, in which three, rather than the usual two, fission products are formed. The third fission product is often tritium, and since its molecular size is very small, it can diffuse through the canning material into the coolant circuit. It emits beta radiation and has a radioactive half-life of 12.6 years. Its biological half-life is around 12 days.

Discharge of fission products into the environment is very strictly controlled, and the authorized release rates for specific isotopes are calculated on the basis of the permissible dose to individuals, which is, of course, well below that which might cause any significant health effect.

Nuclear reactions also produce heavy elements (actinides) whose atomic weight is equal to or higher than that of the uranium isotope from which they are formed. Examples of these actinide elements are the plutonium isotopes, the most important of which is Pu^{239}, a major fissile material. Other plutonium isotopes formed by neutron capture are Pu^{240}, Pu^{241}, and Pu^{242}. Other actinide elements formed in the nuclear reaction include americium-241, americium-243, and curium-244. The actinide elements are important in nuclear waste because of their relatively long half-lives, ranging from 17 years for Cm^{244} to 25,000 years for Pu^{239}. Thus, these actinides require long-term storage, on time scales of 1000 years or more, after discharge from the reactor (see Figure 8.2).

It is often asked: At what time may radioactive waste products be considered safe? A common answer is that the products may be assumed to be safe when their toxic hazard is comparable to that of the original ore from which the fuel was derived and, ultimately, the wastes were generated. A plot of the ratio of the hazard of radioactive waste to that of the original ore is shown in Figure 8.2 for several cases:

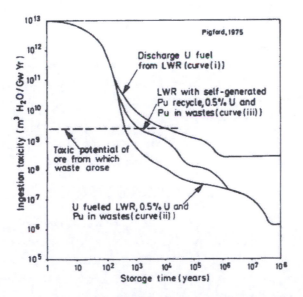

Figure 8.2: Ingestion toxicity of high-level wastes from LWR with and without reprocessing.

1. Fuel discharged from a light-water reactor without reprocessing. Here, the hazard of the waste falls below that of the original ore after 10,000 years.

2. Waste arising from normal reprocessing in which 0.5% of the uranium and plutonium are assumed to be contained in the waste. The hazard level falls to around that of the original ore after about 500 years. It is assumed that 99.5% of the plutonium extracted is used in fast reactors and that the fuel from these reactors would be reprocessed, giving a somewhat similar curve in terms of hazard as a function of time.

3. Waste from thermal reactor fuel in which the plutonium from the reprocessing plant has been incorporated. The hazard is intermediate between those in the first two cases above and falls to that of the original ore after about 1000 years.

Figure 8.2 shows that the hazard falls rapidly after about 100 years for all the cases, reflecting the decay of significant amounts of the shorter-lived fission products. The heat generation rate from the waste products (Figure 7.5) follows curves similar to those in Figure 8.2.

We see, therefore, that after an extended period of time the hazard level of the waste from a given reactor will fall below that of the natural ore sources from which the reactor fuel was derived. Thus the nuclear program would, in the long run (after the cessation of operation of fission power plants), marginally reduce the amount of radioactivity on the earth. However, we must face the need for safe isolation of the waste products during their highly active initial phase, lasting about 1000 years. On a geological time scale, these periods are very short and would not present any difficulty provided care was taken in the placement of the material. In Chapter 1 we discussed the naturally occurring reactor at Oklo. In that case, some of the fission products stayed in the vicinity of the reactor and did not migrate away from it, even though no special precautions were taken to contain them.

8.4 OPTIONS FOR NUCLEAR WASTE DISPOSAL

As we saw above, the most important source of radioactive waste products is the fuel itself. We can illustrate the fuel cycle for a typical thermal reactor as shown in Figure 8.3. Essentially, there are two alternative routes for dealing with the spent fuel. In route A the fuel is passed through a reprocessing plant, which allows recycling of the plutonium and uranium and produces a highly active liquid waste stream. This latter stream may be passed to an interim liquid

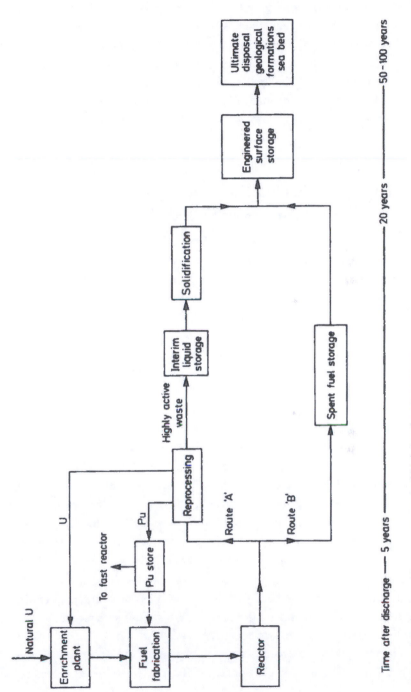

Natural U

Enrichment plant → Fuel fabrication → Reactor

U

To fast reactor ← Pu store

Pu

Reprocessing

Pu store ⇢ Fuel fabrication

Highly active waste

Interim liquid storage → Solidification

Route 'A'

Route 'B'

Spent fuel storage

Engineered surface storage → Ultimate disposal geological formations sea bed

Time after discharge —— 5 years —————— 20 years —————— 50 - 100 years

Figure 8.3: Options for management of high-level active wastes.

storage stage, followed by solidification in one form or another, before being passed to an *engineered surface store*, where it is kept for about 50 years. Ultimately the solidified waste material would be disposed of in suitable geological formations as discussed in Section 8.5. In route B the spent fuel is stored either at the reactor site or away from it in a specially engineered spent fuel store, before ultimate geological disposal. As shown in Figure 8.2, the option of not reprocessing fuel leads to a much longer period before the waste reaches the same level of hazard as the original uranium ore. The options presented by routes A and B are discussed in the next two sections.

8.5 LONG-TERM STORAGE AND DISPOSAL OF SPENT NUCLEAR FUEL

As we saw in Chapter 7, the spent fuel from the reactor may be stored underwater in cooling ponds for typically 10 years or more. The current stage of development of the nuclear program in Europe, the United States, and Japan is such that final decisions about the next phase of spent fuel management, namely *engineered surface storage*, need to be taken in the next few years.

It would be possible, for instance, to continue with away-from-reactor underwater storage, perhaps with the fuel contained in an additional "bottle" to prevent the spread of contamination within such a large water basin store.

Alternatively, a dry storage system could be adopted. Essentially, two different dry storage systems have been developed: the cask or container system and the modular vault dry store. In Canada development of dry store containers for CANDU fuel has been in process for 20 years. The latest design of dry storage container is shown in Figure 8.4. It consists of a box 2.5 m x 2 m x 3.5 m high, constructed from inner and outer steel shells filled with heavy concrete. It weighs 53 tons and contains some 384 CANDU spent fuel bundles: total mass, 7.3 tons. These are stored horizontally in four racks. The Canadian nuclear station at Pickering near Toronto will ultimately have 700 such containers storing nearly 5000 tons of fuel, making it, when complete the world's largest dry store. For more highly rated PWR spent fuel Germany has developed a container using ductile cast iron (CASTOR) 2.4 m diameter. x 4.8 m high, weighing 100 tons when loaded and containing either 33 PWR spent fuel elements or 74 BWR spent fuel elements—15 tons of spent fuel.

An alternative dry storage system is the modular vault dry store illustrated in Figure 8.5. In this concept the spent fuel is contained in individual vertical

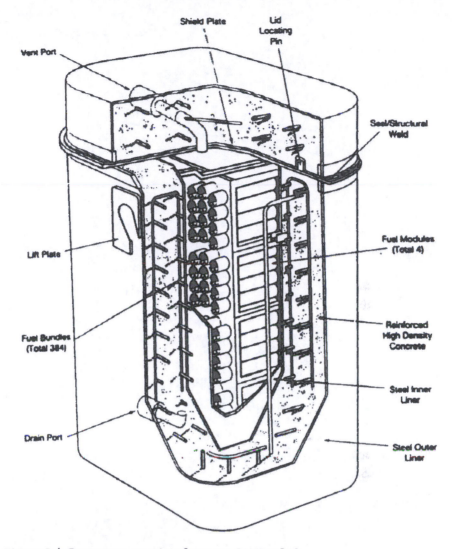

Figure 8.4: Dry storage container for spent CANDU fuel.

sealed fuel storage tubes retained within a concrete vault that can be constructed in modules. Air is drawn in by natural circulation between the array of storage tubes and is discharged via an outlet duct. The coolant air does not come in contact with the spent fuel and therefore neither it nor the concrete structure becomes contaminated. A facility of this type has been constructed to store the spent fuel from the gas-cooled reactor at Fort St. Vrain. This particular

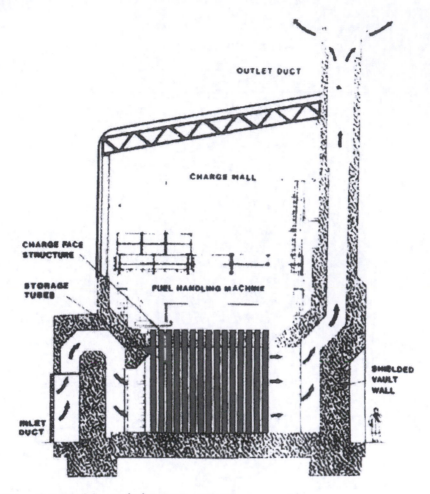

Figure 8.5: Modular vault dry storage system.

design has 45 fuel storage tubes, each holding six fuel elements. The fuel is
moved in and out of the storage tubes by a fuel handling machine moved by
the building crane. A modular vault dry store has been considered by Scottish
Nuclear for dry storage of AGR fuel. Designs for both an 800-ton and a 1200-ton
capacity have been prepared.

Storage in these dry stores would continue for 50–100 years, during which
the level of radioactivity gradually decays (Figure 8.2), as does the rate of heat
production (Figure 7.5). Surface storage in this form for extended periods is advantageous since natural convection cooling can be arranged and the packages

monitored systematically. Ultimately, the rate of heat generation will become low enough to permit storage without special arrangements for natural convection cooling. At this stage long-term disposal may be considered.

The concepts being considered for ultimate disposal of spent or unreprocessed fuel include disposal to underground salt formations or within hard rock geological formations.

8.5.1   Ultimate Disposal in Salt Deposits

Salt deposits are attractive sites for long-term disposal of radioactive waste. The fact that salt is present in the solid form in a geological stratum indicates that it has been free from circulating groundwater since its formation several hundred million years ago. Thus, fuel placed in such a deposit would be free from the leaching action of the groundwater. Salt deposits of this type are quite common, particularly in the United States, and Figure 8.6 shows a conceptual scheme for ultimate disposal of radioactive waste in a salt stratum. Typically, a PWR fuel element may be generating 500 watts of decay heat after 10 years, and this heat generation declines with a half-life of about 30 years since the heat release is dominated by the strontium and caesium decays mentioned above (see Figure 8.3). Thus, after 30 years, the heat release would be down to about 250 watts, and after 60 years it would be reduced to about 120 watts. At these levels, conduction to the surrounding salt strata is sufficient to remove the heat while maintaining the outside surface of the containment canister to a temperature no higher than 100–150°C.

8.5.2   Geological Storage

Geological storage involves the placement of the canisters containing spent fuel elements in a stable stratum typically 1 km below the surface. Such rocks can be assumed to contain water, since the depth would be well below the water table. However, the water is not expected to play a large role in the heat transfer from the blocks, and the store would be designed to maintain the surface temperature of the canisters at no more than 100°C or so. However, the presence of groundwater means that material that is leached from the storage blocks may be transported through the stratum in the water, and this is an important consideration in the design of such systems. Circulation of water through the rock as a result of density differences induced by temperature gra-

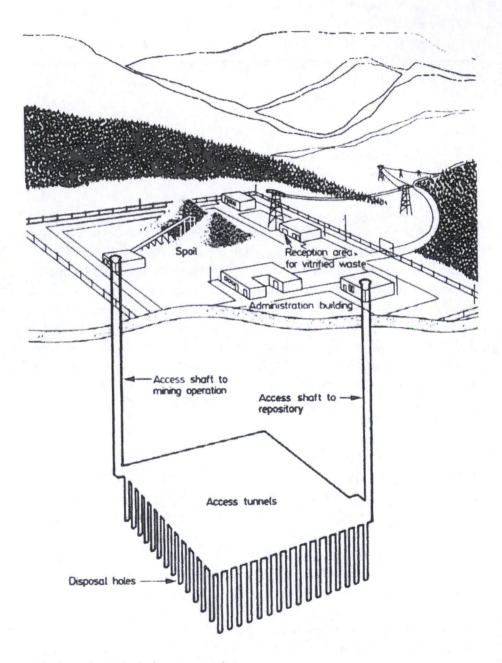

Spoil

Reception area for vitrified waste

Administration building

Access shaft to mining operation

Access shaft to repository

Access tunnels

Disposal holes

Figure 8.6: Geological waste repository.

dients over long periods (the *thermal buoyancy* or *thermal circulation* effect) is important in determining the migration of the fission products. This is a very slow process and is not expected to present a serious hazard, but it must be very carefully taken into account for long-term disposal systems. We discuss such systems further in considering the disposal of fission products from reprocessing plants in the next section.

The choice among the various methods of disposal will be dictated by the availability of suitable storage sites. More geological data will be required before optimum choices can be made. However, studies in many countries indicate that spent fuel can be managed and disposed of without undue risk to humans or the environment.

8.6 STORAGE AND DISPOSAL OF FISSION PRODUCTS FROM REPROCESSING PLANTS

As mentioned in Chapter 7, the nitric acid stream containing the fission products after solvent extraction in the reprocessing plant is concentrated by evaporation and then held in storage tanks. A photograph of one of these tanks under construction is shown in Figure 8.7. Nearly all the high-level waste from the nuclear work in the United Kingdom, accumulated over the past 25 years, is stored in 15 such tanks at Sellafield in Cumbria, which contain a total of about 1000m^3 of liquid.

The stainless steel tanks are contained in concrete vaults, which are themselves lined with stainless steel to provide further containment in the improbable event that the primary container should fail. The space between the tanks and the vaults is monitored, and provision is made for transferring the contents to spare tanks should the need arise. Heat is removed by several independent sets of cooling coils. Reinforced concrete, typically 2 m thick, in which the tanks are sited, protect the operators from direct radiation. Provided cooling is maintained, there are essentially no radiological hazards. The possibility and consequences of an accidental loss of coolant were considered at the Public Enquiry on Windscale in 1978. In the extremely unlikely event of a total loss of coolant (estimated to have a probability of occurrence of 1 in 1 million for each year of operation), it would take hours for the contents to boil and days for them to evaporate, allowing ample time to take remedial action. During the period in which the fission products are generating significant quantities of heat, keeping them in a liquid form facilitates cooling. However, for long-term storage it is considered preferable to convert the waste into solid form, and a number of

Figure 8.7: Cooling coils being inserted into a new high-level liquid waste storage tank at Windscale.

processes have been considered for this.

Work on solidification of nuclear waste started in the 1950s, and by the mid-1960s incorporation of wastes into glass (vitrification) was established on a laboratory scale. The method has been used on an industrial scale in France for a number of years, and the French AVM process (illustrated in Figure 8.8) has been adopted in other countries, including the United Kingdom. Among the alternative processes being investigated is the microwave vitrification process illustrated in Figure 8.9. A range of glass compositions have been developed that enable the constituents of the waste to be incorporated. The glasses have been shown to survive the effects of heating and radiation from the wastes without significant deterioration. They would dissolve very slowly over many thousands of years in freely flowing water. Dissolution in the sort of repositories likely to be used, where access to water is severely restricted, would be very much slower. Other solidification techniques include incorporation into various ceramics and forms of crystalline rocks.

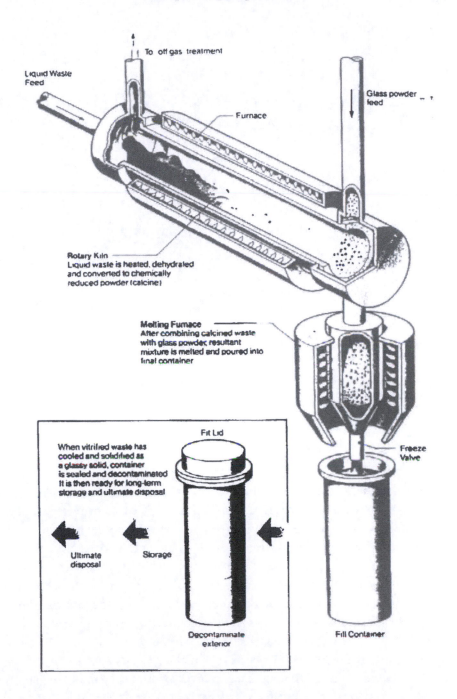

To off gas treatment

Liquid Waste
Feed

Furnace

Glass powder
feed

Rotary Kiln
Liquid waste is heated, dehydrated
and converted to chemically
reduced powder (calcine)

Melting Furnace
After combining calcined waste
with glass powder resultant
mixture is melted and poured into
final container

Freeze
Valve

Fit Lid

When vitrified waste has
cooled and solidified as
a glassy solid, container
is sealed and decontaminated.
It is then ready for long-term
storage and ultimate disposal

Ultimate
disposal

Storage

Decontaminate
exterior

Fill Container

Figure 8.8: French AVM process for vitrification of nuclear waste.

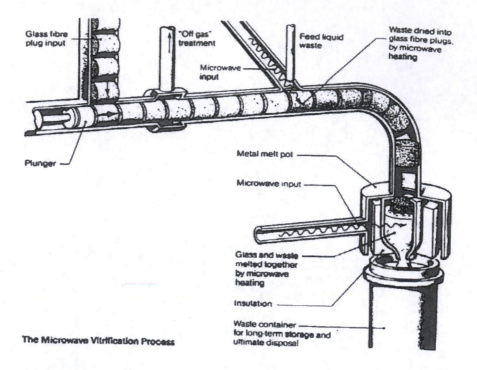

Glass fibre plug input

"Off gas" treatment

Microwave input

Feed liquid waste

Waste dried into glass fibre plugs, by microwave heating

Plunger

Metal melt pot

Microwave input

Glass and waste melted together by microwave heating

Insulation

Waste container for long-term storage and ultimate disposal

The Microwave Vitrification Process

Figure 8.9: Experimental microwave vitrification process. (U.K. Atomic Energy Authority.)

The vitrified waste is typically cast into stainless steel canisters and these canisters dry-stored in a manner illustrated in Figure 8.5. The vitrified waste canisters will be stored in these natural convection air-cooled stores up to 50 years before final disposal. A typical glass block might be 30 cm in diameter and 1 m long, weighing about 0.2 tons. About 20% of the weight of such a block would be the fission products from the reprocessing plant, the rest being added materials to help form the glass. As in the case of the spent fuel, the heat release is dominated by the caesium-strontium decay with a half-life of 30 years. It is generally considered that surface temperatures for the block in the long-term store should fall below 100°C, and at this surface temperature a heat rejection rate of about 1 kW is achievable by conduction into the surrounding rocks. To avoid interactions between blocks within the rock matrix, a spacing of about 10 m in all directions is required. This could be achieved by tunneling to the required 1 km depth and then constructing a gallery from which holes, say, 200 m in depth, are drilled. The blocks could then be dropped in and the required 10 m spacing achieved by infilling before dropping in the next block. The holes themselves would also be spaced out on an array of 10 m square.

For the British program, it is estimated that some 10,000 blocks will have been produced before the end of the century. This would imply the use of an array of around 50 x 50 blocks arranged, say, in a cube.

Again, the problem of leaching of fission products from the block and their transfer through the strata must be considered, and the thermal circulation and thermal buoyancy effects mentioned are very important in the medium term. Enough is now known about these systems to be sure that safe disposal of nuclear waste is possible.

8.7 DISPOSAL OF OTHER MATERIALS

As we saw in Section 8.1, a large variety of low-level wastes also arises from the nuclear program. Waste consisting of miscellaneous rubbish (such as rubber gloves and tissues contaminated with traces of radioactive material) is typically contained in steel drums, compacted to reduce bulk and then placed in steel containers and disposed of in a shallow trench area covered with at least 1 m of soil. Measurements on such an engineered disposal facility have indicated that the radiological significance of the disposal is negligible.

Wastes with medium levels of radioactivity from reprocessing, power reactor operations, and decommissioning, as well as plutonium-contaminated wastes, are usually contained within a concrete or bitumen matrix within stainless steel drums. There is widespread agreement that geological disposal is the best solution for the management of such wastes. A geological repository of the type illustrated in Figure 8.6 would be suitable. Sweden already has an operational repository under the seabed for medium- and low-activity wastes at Forsmark; Finland has a similar repository at it Olkiluoto power station site. These facilities are about 100 m below the seabed-ground level. In Britain, UK Nirex Ltd. was set up in 1985 with the responsibility of providing radioactive waste disposal facilities. Currently, a deep underground site near Sellafield, northwest England, is being investigated for a geological repository. An underground rock laboratory is planned as the first stage prior to the construction of the facility, expected to be brought into operation around 2010.

Liquid wastes at low activities arise from all nuclear sites, particularly from reprocessing plants, and are discharged under the regulations laid down by the licensing authority. Obviously, great care must be taken to avoid any public danger from such discharges.

Gaseous wastes, typically noble gas isotopes, are also produced from reactors and reprocessing plants. These are normally discharged to the atmosphere under carefully controlled conditions.

A final point on disposal concerns the decommissioning of a nuclear plant. Decommissioning is done in stages; stage 1 is concerned with the removal of spent fuel—*defueling*—from the reactor. This starts at shutdown and can take up to 3 years for a large gas-cooled reactor. The spent fuel that is discharged is then managed in the same way as "operational" spent fuel. This reduces the total amount of radioactivity at the reactor site to less than one-seventh that at shutdown. The second stage involves all the dismantling of all nonradioactive plant and buildings other than the reactor and its concrete biological shield. This stage follows on from stage 1 and takes 5 to 10 years. The reactor building itself is then sealed for a period of surveillance. Finally, stage 3 involves the complete dismantling of the reactor and returning the site to a "greenfield" status. This stage occurs about 100 years after shutdown and takes about 10 years to complete.

A variant on this strategy involves the construction of a high-integrity intruder-proof containment around the reactor building—*Safestore*—that can be left for periods of up to 100 years before the final dismantling of the reactor. This strategy allows the maximum time for the radioactivity in the reactor building to decay, thus minimizing the hazard when actual dismantling takes place. Modern PWR stations are designed for the replacement of all components with the exception of the reactor pressure vessel, and are therefore relatively straightforward to decommission.

So far about 80 nuclear reactors have been shut down worldwide and several sites have been cleared completely—the world's first civil PWR station, Shippingport, for example. In the United Kingdom, decommissioning has started at three of the older Magnox station sites, Berkeley, Hunterston, and Trawsfynydd. Handling and disposal of radioactive waste from decommissioning follow similar routes to reprocessing and reactor operational wastes. Decommissioning represents only a small fraction (approximately 5% maximum) of nuclear generating costs.

REFERENCES

Cooper, J.R., and J.W. Rose (1977). *Technical Data on Fuel*, p. 53. Scottish Academic Press.

Ealing, C.J. (1994). "Experience and Application of the GEC Alsthom Modular Vault Dry Store." *Nuclear Engineer* 35 (March–April): 48–54.

Janbury, K. (1994). "Transport, Storage and Final Disposal of Spent Fuel in the Federal Republic of Germany." *Nuclear Engineer* 35 (May–June): 78-83.

OECD (1988). *Environmental Impacts of Renewable Energy*. Report by the Organization for Economic Cooperation and Development.

Passant, F.H. (1994). "Waste Management and Decommissioning." *Nuclear Energy* 33 (4): 223–229.

Stevens-Guille, P.D., and F.E. Pave (1994). "Development and Prospects of Canadian Technology for Dry Storage of Used Nuclear Fuel." *Nuclear Engineer* 35 (March–April): 64–71.

EXAMPLES AND PROBLEMS

1 Heat fluxes in sealed storage casks

Example: For the sealed storage cask concept illustrated in Figure 8.4, assume a simplified model of the heat transfer behavior in which heat is released from the 19-in. (0.48-m) diameter carbon steel flask by convective heat transfer to an air stream flowing up the annular gap at a rate of 0.2832 m^3/s (600 ft^3/min) and by radiation to the inner surface of the 31-in. (0.7874 m) inside diameter concrete gamma/neutron shield. Any heat radiated to the shield inner surface is assumed to be removed by convection to the air stream. Calculate the heat removal rate by convection from both surfaces of the annular gap and compare the results with the stated heat release rate (5 kW) from the stored element. Assuming an emissivity of 0.82 for the steel cask, what value of emissivity would need to be assigned to the inner surface of the neutron shield to be consistent with this simplified picture? For the calculations, assume the following physical properties for the air flowing by natural circulation through the gap, appropriate to the mean air temperature in the gap, i.e., 35°C (95°F):

Density (ϱ)	1.146 kg/m^3
Viscosity (μ)	1.83×10^{-5} kg/ms
Specific heat capacity(c_p)	1025.8 J/kg K
Thermal conducivity (k)	2.68×10^{-2} W/m K

Assume that the effective length of the surfaces in the axial direction is 3.2 m (i.e., 10.5 ft, the length of the cask) and ignore end effects.

Solution: The first step is to calculate the velocity V_{sp} of the air in the channel. The volumetric flow rate is 0.2832 m^3/s (600 ft^3/min), and the cross-sectional area of the annular gap is given by

$$A = \frac{\pi}{4}\left(D_o^2 - D_i^2\right)$$

where D_i is the inside diameter and D_o the outside diameter. Substituting $D_o = 0.7874$ m (31 in.) and $D_i = 0.4826$ m (19 in.), we have $A = 0.3040$ m^2, and the velocity is given by

$$V = \frac{\text{volumetric flow rate}}{A} = \frac{0.2832}{0.3040}$$
$$= 0.9315 \text{ m/s (3.06 ft/s)}$$

The convective heat transfer coefficient can be calculated for the airflow from the (Dittus-Boelter) relationship:

$$\text{St} = 0.023 \, \text{Re}^{-0.2} \, \text{Pr}^{-0.6}$$

where

$$\text{St} = \frac{\alpha}{\varrho V c_p} = \text{Stanton number}$$

$$\text{Re} = \frac{\varrho V D_e}{\mu} = \text{Reynolds number}$$

$$\text{Pr} = \frac{c_p \mu}{k} = \text{Prandtl number}$$

where α is the heat transfer coefficient and D_e is an equivalent diameter that is evaluated for the annulus from the expression

$$D_e = D_o - D_i = 0.7874- = 0.4826 = 0.3048 \text{ m}$$

We first evaluate Re and Pr as follows:

$$\text{Re} = \frac{\varrho V D_e}{\mu} = \frac{1.146 \times 0.9315 \times 0.3048}{1.83 \times 10^{-5}}$$
$$= 1.778 \times 10^4$$
$$\text{Pr} = \frac{c_p \mu}{k} = \frac{1025.8 \times 1.83 \times 10^{-5}}{2.68 \times 10^{-2}} = 0.7005$$

Thus, the Stanton number is given by

$$\text{St} = 0.023 \, \text{Re}^{-0.2} \, \text{Pr}^{-0.6}$$
$$= 0.023 \times (1.778 \times 10^4)^{-0.2} (0.7005)^{-0.6}$$
$$= 0.023 \times 0.1413 \times 1.238$$
$$= 4.023 \times 10^{-3}$$

The heat transfer coefficient α is given by

$$\alpha = \text{St } \varrho V c_p$$
$$= 4.023 \times 10^{-3} \times 1.146 \times 0.9314 \times 1025.8$$
$$= 4.405 \text{ W/m}^2\text{K}$$

The heat transfer rates $\dot{Q}_i$ and $\dot{Q}_o$ from the inner and outer surfaces can be calculated from the relationships:

$$\dot{Q}_i = A_i \alpha (T_i - T_a)$$
$$\dot{Q}_o = A_o \alpha (T_o - T_a)$$

where A_i and A_o are the surface areas of the respective surfaces, T_i and T_o the surface temperatures, and T_a the air temperature. From Figure 8.4 we have T_i = 360°F = 182.22°C = 455.37 K, and T_o = 210°F = 98.89°C 372.04 K. The mean air temperature T_a is 35°C (95°F), and the areas are given by $A_i = \pi D_i L$ and $A_o = \pi D_o L$, where L is the total length of the surface. Thus:

$$A_i = \pi \times 0.4826 \times 3.20 = 4.85 \text{ m}^2$$
$$A_o = \pi \times 0.7874 \times 3.20 = 7.92 \text{ m}^2$$
$$\dot{Q}_i = A_i \alpha (T_i - T_a)$$
$$= 4.85 \times 4.405(182.22 - 35)$$
$$= 3145 \text{ W}$$
$$\dot{Q}_o = A_o \alpha (T_o - T_a)$$
$$= 7.92 \times 4.405(98.89 - 35)$$
$$= 2229 \text{ W}$$

The total calculated heat transfer rate is $\dot{Q}_i + \dot{Q}_o$ = 5374 W = 5.374 kW, which is in reasonable agreement with the stated figure of 5 kW (certainly within the accuracy of the heat transfer coefficient calculation method).

The rate of heat transmission $\dot{q}_R$ from a surface by thermal radiation is given by

$$\dot{q}_R = \varepsilon \sigma T^4$$

where ε is the emissivity, σ the Stefan-Boltzmann constant (5.6696 x 10^{-8} W/K⁴ M²), and Tsp the absolute temperature (in kelvins). For our simple model, assume that all heat radiated from the inner surface reaches the outer surface and vice versa. Thus, the net rate of transfer from the inner to the outer surface is given by

$$\dot{Q}_{Rio} = \dot{q}_{Ri} A_i - \dot{q}_{Ro} A_o$$
$$= \varepsilon_i \sigma T_i^4 A_i - \varepsilon_o \sigma T_o^4 A_o$$
$$= 2229 \text{ W (i.e., equal to } \dot{Q}_o \text{ from the convective heat transfer calculation)}$$

Thus, to calculate ε_o as required, we can rearrange the above formula:

$$\varepsilon_o = \frac{\varepsilon_i \sigma T_i^4 A_i - 2229}{\sigma T_o^4 A_o}$$

$$= \frac{0.82 \times 5.6696 \times^{-8} \times 455.37^4 \times 4.85 - 2229}{5.6696 \times 10^{-8} \times 372.04^4 \times 7.92}$$

$$= \frac{9695 - 2229}{8602} = 0.87$$

This value of emissivity is reasonably consistent with the range of values for a concrete surface. However, a more complete model of the system should take account of conduction through the shield, heat generation in the shield, and heat losses from the outside of the shield.

Problem: Suppose that the air ingress into the storage unit investigated in the example is partially blocked in a manner that results in the airflow rate being reduced by 50% (down to 0.416 m³/s). What would the consequences be, in terms of the simple model used in the example, for the convective and radiant fluxes and the canister wall and shield inner wall temperatures?

2 Cooling of a high-level liquid waste storage vessel

Example: High-level liquid waste is stored in a vessel whose diameter D is 6 m. The liquid level h in the vessel is 5 m, and the fission product heat is removed by cooling with water that passes through coils of 5-cm-outside-diameter stainless steel tube, the coils being immersed in the waste liquid (see Figure 8.7). The water enters the coils at 20°C and leaves at 25°C. The liquid waste is generating 14 kW/m³ of fission product decay heat and must be maintained at a temperature less than 35°C to minimize corrosion. Calculate the water flow rate needed to maintain cooling and the length of tube required in the coils, assuming an overall heat transfer coefficient U of 350 W/m² K. Assume a specific heat c_p for water of 4.18 kJ/kg K.

Solution: The volume V of waste liquid in the vessel (ignoring the volume occupied by the coils) is given by

$$V = \frac{\pi D^2 h}{4} = \frac{\pi \times 6^2 \times 5}{4} = 141.4 \text{ m}^3$$

The heat generation rate $\dot{Q}$ is thus 141.4 x 14 = 1.98 MW. The cooling water mass $\dot{M}$ required is given by

$$\dot{M} = \frac{\dot{Q}}{c_p \Delta T_{ol}}$$

where ΔT_{ol} is the temperature difference between the cooling water outlet and inlet. Thus,

$$\dot{M} = \frac{1.98 \times 10^6}{4.18 \times 10^3 \times (25 - 20)}$$

$$= 94.74 \text{ kg/s}$$

If A is the surface area of the outside of the coil tubes,

$$\dot{Q} = UA\Delta T_m$$

when ΔT_m is the mean temperature difference between the cooling water and the liquid waste. This is given by

$$\Delta T_m = 35 - \left(\frac{20 + 25}{2}\right) = 12.5°C$$

and hence

$$A = \frac{\dot{Q}}{U\Delta T_m}$$
$$= \frac{1.98 \times 10^6}{350 \times 12.5}$$
$$= 452 \text{ m}^2$$

The surface area per meter of tube is $\pi \times 0.05 = 0.157$ m^2, and hence 452/0.157=2879 m of coil tubing is required.

Problem: For the tank described above, calculate the volume occupied by the coils, and correct the calculations to take account of this volume. Also calculate the consequences of a cooling-water failure in half of the coils.

3 Heat dissipation from buried waste blocks

Example: Active waste from a reprocessing plant is vitrified into cylinders with a diameter D of 0.3 m, which are buried in a trough (the cylinders being end to end) underground at a depth x of 7 m. The cylinders emit I kW/m fission product decay heat. Calculate the surface temperature T_1 of the cylinders, assuming that the soil surface temperature T_2 is 20°C and the soil has a thermal conductivity of 1 W/m K.

Solution: The rate of heat transmission from a buried cylinder is of considerable interest with respect to pipes buried in the ground and is given by (Cooper and Rose, 1977):

$$\dot{Q} = \frac{2\pi k (T_1 - T_2)}{\ln\{2x / D + [(2x / D)^2 - 1]\frac{1}{2}\}}$$

where Q is the amount of heat transmitted per unit length (1 kW/m in the present example). Substituting $k = 1$ W/m, $x = 7$, and $D = 0.3$, we have

$$\dot{Q} = 1000 = \frac{2\pi k (T_1 - 20)}{\ln\{2 \times 7 / 0.3 + [(2x / D)^2 - 1\frac{1}{2}\}}$$
$$= \frac{6.283(T_1 - 20)}{4.536}$$

Thus:

$$T_1 = \frac{1000 \times 4.536}{6.283} + 20$$
$$= 742°C$$

Problem: Repeat the above calculation for the case in which the heat release rate from the cylinders is 500 W/m and the cylinders are buried at a depth of 4 m.

BIBLIOGRAPHY

Allday, C. (1982). "Nuclear Fuels: Development, Processing and Disposal." *Energy World* (95): 12–15.

———. (1982). "BNFL's Reprocessing Work and Experience: Past, Present and Future." *Nucl. Eur.* 11 (6): 11–13.

Black, J.H., and N.A. Chapman, (1981). "In Search of Nuclear Burial Grounds." *New Sci.* 91 (1266): 402–404.

Bradley, N., and G.A. Brown (1981). "Natural Draught Centralized Dry Store for Irradiated Fuel and Active Waste." *Nucl. Eng. Int.* 26 (320): 38–40.

Celland, D.W., and A.D.W. Corbet (1982). "Vitrifying Britain's Waste." *Nucl. Eng. Int.* 26 (331): 33–35.

Deacon, D. (1981). "The Long-Term Dry Storage of Irradiated Oxide Fuel and Vitrified Waste." *Nucl. Eng. Int.* 26 (317): 32–36.

Grover, J.R. (1979). "High-Level Waste Solidification—Why We Chose Glass." Report AERE-R-9432, UKAEA Harwell, 14 pp.

Lewis, J.B. (1983). "The Case for Deep Sea Disposal." *Atom* (317): 49–52.

Radioactive Waste Management (1982). Government White Paper updating the position on radioactive waste management from that covered in the 1977 White Paper, *Nuclear Power and the Environment.* HMSO, London, Cmnd. 8607, 20 pp.

9

Fusion Energy
Prospect for the Future

9.1 INTRODUCTION

In the preceding chapters we have seen how uranium, mined from the earth's crust, is utilized in a nuclear reactor to create energy and how the resulting waste products can be dealt with safely. We have concentrated on the thermal or heat-generating aspects of the materials at the various stages of the cycle. We have seen that the energy that can be recovered from nuclear fission of 1 ton of uranium can be increased 60-fold by the use of fast reactors and that this can extend our use of fission power from a few tens to many hundreds of years. Nevertheless, the world's uranium resources are finite, and energy resources will increasingly be required by the developing world. Scientists have therefore turned to alternative ways to release nuclear energy. What more natural place to look than to the ultimate source of the earth's energy—the sun. The energy generated by the sun is not the result of splitting up nuclei of heavy elements but of the joining together—*fusion*—of nuclei of light elements such as the isotopes of hydrogen or lithium. These elements are abundant and easily available on the earth, so what is the problem of releasing fusion energy for our use?

The problem is that to release the energy of fusion in a controlled manner requires heating the reacting nuclei to temperatures of tens to hundreds of millions of degrees and holding them in sufficient quantities at these temperatures long enough for the reaction to take place. A device capable of creating such a reaction is called a *thermonuclear reactor.*

The energy release in the sun results from the conversion of hydrogen into helium. Effectively four protons fuse together to form one helium nucleus with an energy release of 7.7 x 10^{-13} joules. Thus the conversion of 1 gram of hydrogen to helium produces 0.71 x 10^{12} joules. The energy released by the sun is almost incomprehensibly large: 0.39 million million gigawatts (3.9 x 10^{26} watts). This requires the consumption of 5.5 x 10^{14} grams/s. (or alternatively, 550 million tons per second). Even so, the sun has an expected lifetime of 10,000 million years!

On Earth it is not possible to reproduce the solar conditions. The specific thermonuclear reaction is too slow to produce a practical size of reactor. Fortunately there are other fusion reactions that might form the basis of a practical reactor.

9.2   THE FUSION PROCESS

Faster fusion reactions are possible with a range of mixtures involving the isotopes of hydrogen, helium, and lithium. These include:

$$^2D + {}^2D \rightarrow {}^2He + n + 0.96 \times 10^{-13} \text{ J}$$
$$^2D + {}^2D \rightarrow {}^2T + {}^1H + 1.19 \times 10^{-13} \text{ J}$$
$$^2D + {}^3D \rightarrow {}^4He + n + 5.2 \times 10^{-13} \text{ J}$$

Most research effort is being directed at the last of the reactions because it is the least difficult reaction to achieve (Figure 9.1).

Most (80%) of the energy released is in the form of kinetic energy of the neutron. Note that though the energy released per fusion reaction is typically 10 times less than for a single fission reaction, the neutrons are released with perhaps 5 times as much energy.

Deuterium, as we saw in Chapter 3, occurs in ordinary water at a concentration of 0.016% and can be readily separated by chemical processes. Tritium

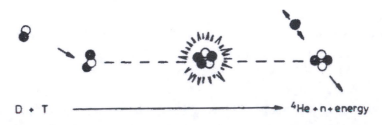

D + T ⟶ ⁴He + n + energy

Figure 9.1: Deuterium-tritium fusion reaction.

does not occur naturally but can be produced from lithium by bombardment with a neutron. Thus:

$$^6\text{Li} + \text{n} \rightarrow\ ^4\text{He} + ^3\text{T} + 7.7 \times 10^{-13}\ \text{J}$$
$$^7\text{Li} + \text{n} \rightarrow\ ^4\text{He} + ^3\text{T} + \text{n} - 4.0 \times 10^{-13}\ \text{J}$$

As we shall see later, it is possible to arrange the system so that the neutrons from the fusion reaction are used to breed more tritium from these reactions. This is done in a *blanket* in a manner similar to that used in a fast fission reactor. The power generated in such a fuel cycle for each gram of lithium is 36 million joules (10,000 kWh).

It is interesting to compare the energy available from these isotopes with, say, the figures given in Table 1.2. There, we saw that the world's readily available uranium resources, utilized in fast reactors, could release around 10^{23} J and using the uranium in the ocean could increase this figure to around 10^{26} J. These figures compare with the current world electrical consumption of 1.8 x 10^{19} J/year. In fusion reactions the deuterium in the oceans could release around 3 x 10^{31} J, while the land-based lithium reserves could yield around 10^{28} J, and including the lithium in the oceans would raise the figure to 2 x 10^{28} J. Thus, the fuel resource with fusion reactions can be considered limitless, certainly beyond a million years.

Let us therefore turn our attention to how we might tap into this immense source of energy. The fusion reaction is difficult to achieve because the deuterium and tritium nuclei are each positively charged electrically. Like charges repel each other and this force can be overcome only if the nuclei approach each other with sufficient velocity—millions of kilometers an hour—to overcome the mutual repulsion. That means heating up the gaseous deuterium-tritium mixture to a temperature around 100 million degrees or more. At a temperature of a few thousand degrees the gas becomes ionized; that is, the electrons separate from the atoms and the separate electrons and nuclei move randomly (Figure 9.2). Such a mater-

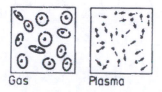

Gas   Plasma

Figure 9.2: Gas and plasma states.

ial is known as a *plasma*. Plasmas exist in the sun and stars and also in such everyday items as neon signs and electric arcs.

It is not enough to heat the plasma to the required temperature. It is also necessary to hold the plasma at that temperature for sufficient time for the reaction to take place. Clearly, the length of time will depend on the number of nuclei in a given volume of plasma. The required conditions have been identified in the Lawson criterion, which states that the product of the time for which the plasma is confined τ_E and the density of the plasma (n) must be greater than 10^{20} s/m^3

$$n \times \tau > 10^{20} \text{ s/m}^2$$

Thus, if the density of the plasma is 10^{20} nuclei per cubic meter, the plasma must be held at 100–200 million degrees for 1 s.

9.3 CONFINEMENT

How are we to "confine" the plasma long enough so that it does not touch (and melt) the walls of the vessel in which the reaction is to take place? In the Sun and stars the fusion plasma is held together by large gravitational forces. On Earth we obviously cannot use such forces to contain a plasma in any convenient-sized apparatus. Two ways have been tried to provide this *confinement* of the plasma.

1. *Magnetic confinement.* Since plasmas are excellent conductors of electricity, they can be acted on by magnetic fields (Figure 9.3). Thus, magnetic fields can be used to shape and confine the plasma in such a manner that it does not touch the walls of the vessel in which the gaseous mixture is held. If the plasma did come into contact with the vessel walls, it would *quench*, losing its energy and high temperature very rapidly.

2. *Inertial confinement.* The alternative to magnetic confinement is to contain the isotopic mixture frozen solid at about 15 K as a small spherical pellet or

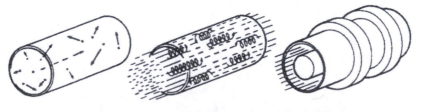

Without magnetic field   Charges in a   Plasma in a
                                magnetic field   magnetic field

Figure 9.3: Confining effect of the magnetic field.

bead (Figure 9.4). This spherical pellet is then bombarded from every direction by beams of high-powered lasers, which compress and heat the mixture to fusion temperatures. Inertia holds it together long enough—perhaps a nanosecond (10^{-9}s) for the fusion reaction to take place. This time can be so short because of the very high densities achieved.

Considerable research is being done on the inertial confinement process at the Lawrence Livermore Laboratory at the University of Rochester, New York and at the Los Alamos Laboratory, New Mexico. There are, however, fundamental difficulties with this route to a practical system. These are the low efficiency of the laser (1–2%), the low fraction of fusion energy released to date (~0.01%), and the difficulties of engineering a device to produce a continuous power output involving ignition of a stream of frozen pellets at a high rate.

Most effort is therefore being devoted to trying to achieve a fusion reaction using magnetic confinement. The structure of magnetic fields is often indicated by *lines of force* or *field lines*; the stronger the field, the greater the density of the lines. Within a magnetic field, charged nuclei take a spiral path in the direction of the field lines as illustrated in Figure 9.5*a*. A magnetic field line causes a charged nucleus to spiral around it (Figure 9.5*a*). If the field is arranged so as to close on itself in a circle within a circular chamber (Figure 9.5*b*), the particles will spiral around the field and remain trapped within the circular chamber, or *torus*. Unfortunately, this does not always happen in practice due to instabilities that occur in the plasma. Nevertheless,

 1 Hollow sphere containing a D & T mixture.

 2 Surface heated by laser radiation

 3 Sphere vaporises and expands inwards and outwards

4 The sudden inwards movement compresses and heats the D & T mixture up to conditions for fusion to take place

 5 Energy released as an explosion liberating neutrons

Figure 9.4: Intertial confinement.

a

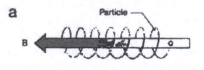

A charged particle will gyrate around a magnetic field line B.
This is the Basic Mechanism whereby a magnetic field confines a plasma.

b

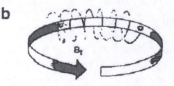

In a Toroidal system the field lines, B_t are bent back on themselves to form a closed loop.

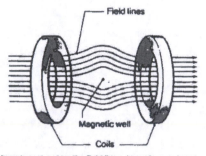

In an 'open' system the field lines do not form a closed loop within the device.

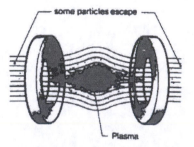

The bulk of the particles form a plasma sitting in the Magnetic Well. Some particles have high longitudinal energy and can escape out the ends.

Figure 9.5: (*a*) Particle spiraling around a magnetic field line. (*b*) A closed toroidal system. (*c*) A magnetic mirror or bottle.

most of the experiments that have tried to achieve controlled fusion reactions make use of this closed doughnut-shape configuration. Another possibility is to constrict the magnetic field lines at each end of a tube. Particles trying to escape by spiraling along the field lines are reflected back into the central region. This arrangement is called a magnetic *mirror* or *bottle* (Figure 9.5*c*).

9.4   CURRENT TECHNICAL POSITION

Research into controlled fusion reactions is proceeding in the United States, Russia, Japan, and Europe. One particular configuration of magnetic fields has proved promising: the so-called Tokamak configuration. For interested readers, Figure 9.6 shows the details of the Tokamak device and explains why three separate magnetic fields are used to control the plasma. Experiments have been conducted with larger and larger devices. Figure 9.7 shows the progress made

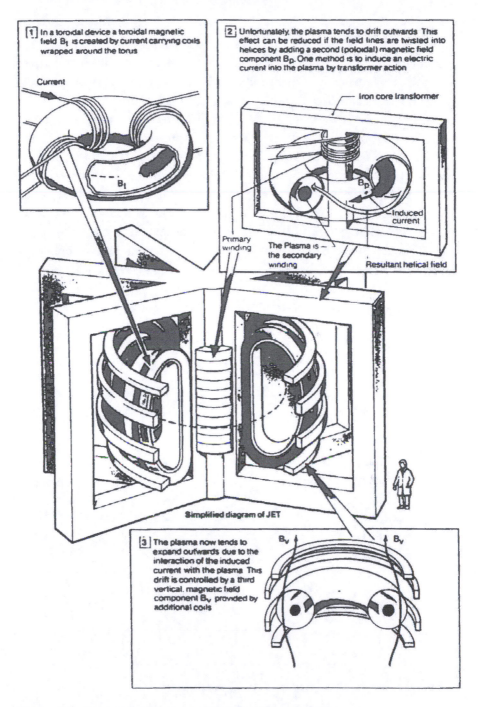

Figure 9.6: Reasons for the appearance of a Tokamak device.

toward the achievement of the Lawson criterion. One particular experimental Tokamak is the so-called JET (Joint European Torus) project. The scale of this experiment can be judged from the fact that the mean diameter of the torus is 6 m and the radius of the plasma will be 1.25 to 2.10 m. An illustration of what JET looks like is shown in Figure 9.8.

On November 9, 1991, at 07.44 P.M., the JET experiment produced about 2 MW of fusion power, the first time that a significant amount of power had been obtained from controlled nuclear fusion reactions. This was also the first time JET had been operated with mixtures of the isotopes (14% tritium–86% deuterium). In practice only about 0.2 gram of tritium was used. For this experiment JET consumed far more power than it generated. The next target is *breakeven*, the production of as much fusion power as consumed in heating the plasma. Further experiments are planned at JET to obtain and study plasmas under conditions and dimensions approaching those needed in a thermonuclear reactor (Figure 9.7).

Plasmas are usually heated by passing a current through the electrically conducting plasma. This form of heating (Figure 9.9) is effective up to about 10 mil-

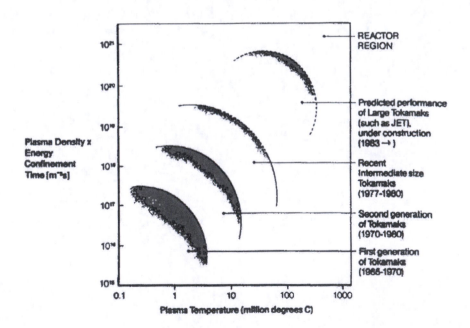

Figure 9.7: Progress toward fusion reaction.

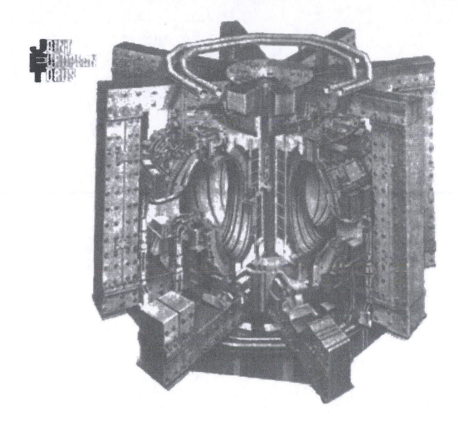

Figure 9.8: Joint European Torus (JET).

lion degrees, but if attempts are made to increase the current, instabilities set in. Other methods of heating include one in which a beam of ionized particles is accelerated up to high energies, neutralized, and fired into the plasma. Once in the plasma, the particles become ionized, are trapped, and transfer their energy by collision with the plasma electrons. Other methods of heating include radio frequency (RF) heating and compression of the plasma with a magnetic field. All these methods have been tried on JET. They are now understood and their use can be contemplated on fusion reactors with confidence.

From this short discussion of the present status of our knowledge of fusion reactions, it will be seen that we have just about reached the point where a controlled fusion reaction has been demonstrated. Thus we have reached the point in the development of fusion power that Fermi achieved in 1942 with the first

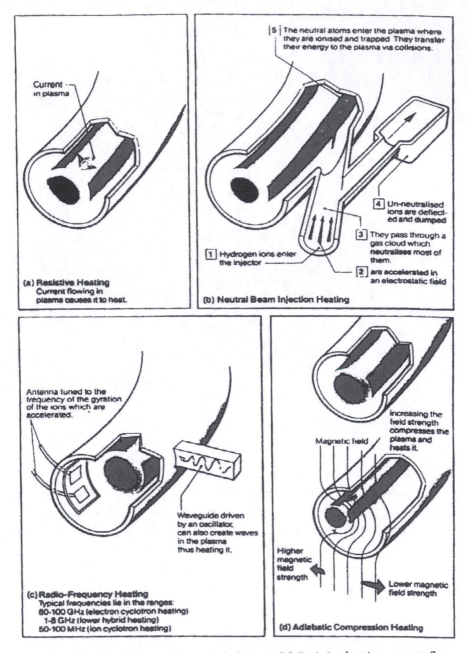

Figure 9.9: Methods of heating torodial plasmas. (*a*) Resistive heating; current flowing in plasma causes it to heat. (*b*) Neutral beam injection heating. (*c*) Radio-frequency heating; typical frequencies lie in ranges: 60–100 GHz (electron cyclotron heating), 1–8 GHz (lower hybrid heating), 50–100 MHz (ion cyclotron heating). (*d*) Adiabatic compression heating.

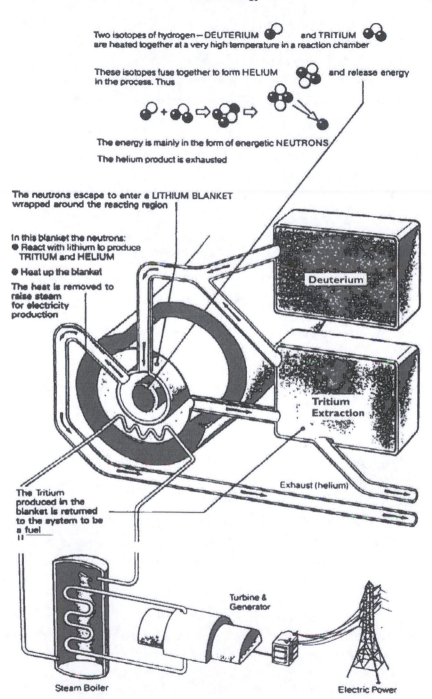

Two isotopes of hydrogen—DEUTERIUM   and TRITIUM
are heated together at a very high temperature in a reaction chamber

These isotopes fuse together to form HELIUM   and release energy
in the process. Thus

The energy is mainly in the form of energetic NEUTRONS

The helium product is exhausted

The neutrons escape to enter a LITHIUM BLANKET
wrapped around the reacting region

In this blanket the neutrons:
● React with lithium to produce
   TRITIUM and HELIUM
● Heat up the blanket

The heat is removed to
raise steam
for electricity
production

Deuterium

Tritium
Extraction

The Tritium
produced in the
blanket is returned
to the system to be
a fuel

Exhaust (helium)

Turbine &
Generator

Steam Boiler

Electric Power

Figure 9.10: Deuterium-tritium fusion.

controlled fission reaction. Progress is limited by (1) the capabilities of removing the heat from the neutrons deposited in the blanket and (2) the damage sustained due to the high-energy neutron radiation of the thin-walled vessel.

To progress to such a reactor requires the solution of many technical and engineering problems, a fair number of which involve the thermal engineer. (See *Nuclear Engineering and Design*, vol. 68, April 1982). For example, various coolants have been suggested to remove the heat from the blanket into heat exchangers to raise steam. It is possible to use the lithium itself, or an alternative liquid metal, although the intense magnetic fields impose a very high pressure drop with consequent high pumping losses. An alternative is to use a gaseous coolant, but this limits the energy density possible through the vacuum vessel wall (*first wall*).

Bickerton (1993) has summarized some of the requirements for the reactor:

- Start-up, ionise gas and increase ring current to final value (~ 20 MA)
- Heat plasma with auxiliary heating, typically < 100 MW to ignition point
- Maintaining ignition, manoeuvre plasma parameters to final operating point, where required fusion thermal power (~3 GW) is achieved
- Maintain plasma current in steady state for long pulse (~ 1000 sec.)
- Extract fusion ash; i.e. helium in the case of D-T reactions
- Refuel plasma (with fresh deuterium and tritium in case of D-T reactions
- Extract heat at high efficiency from blanket, tritium breeding ration > 1.0
- Shield super-conducting coils from neutron induced heating and radiation damage
- Maintain system remotely, e.g. change first wall every 2–5 years (because of neutron damage)
- Decommission and dispose of wastes.

Solutions of these and other engineering problems are being sought in the design of the next generation of Tokamaks. Because of the scale of the experiment, all the major nations involved in fusion research are collaborating on the design of the next machine, ITER (International Tokamak Experimental Reactor).

The design of this machine is based on the scaling laws derived from previous experiments that predicted plasma performance as a function of machine parameters. The main ITER parameters are given as:

Fusion power	1.5 GW
Burn time	1000 sec.
Plasma current	24 MA
Major radius	8.1 m
Plasma radius	3.0 m
Magnetic field	5.7 Tesla

The overall objective is nothing less than a demonstration of the feasibility of fusion energy for peaceful purposes. The outline design is well advanced (Toschi, 1995).

Regarding impact on the environment, fusion reactors have some advantage over fission reactors. The waste product of the fusion reaction, helium, is inert, and thus the problem of managing highly radioactive waste does not arise. The structure of the reactor itself will become intensely radioactive and will require remote maintenance. But this radioactivity will decay over periods of hundreds rather than tens of thousands of years. The tritium used in the reactor represents a radiological hazard, and since it is an isotope of hydrogen, it requires very careful containment and protection against accidents, such as fires. In summary, although the potential radiation hazards presented by fusion reactors will be less than those of fission reactors, they will require careful attention at the engineering design stage.

9.5 CONCLUSIONS

With the probable development of fusion in addition to fission energy, nuclear power presents humans with a virtually infinite source of energy. The central role of energy in our economic structure has been very clearly demonstrated over the 20 years since the oil crises of the 1970s. Nuclear fission energy provides a proven resource for the immediate future and nuclear fusion energy a great potential resource for the more distant future. Humanity must make use of these resources, particularly if the underdeveloped world is to achieve freedom from the bondage of hunger, disease, and poverty, and the world is to sustain its development.

Of course, there are many technical problems still to be solved, and the utilization of nuclear power will demand continual vigilance and great attention to technical detail if it is to continue its very successful beginning. Not least of these problems are those associated with the removal of heat from the nuclear reaction and its effective application in power generation. We thus make no apology for having written this book from our own viewpoints, those of thermal engineers.

In addition, there are great institutional and organizational problems to be properly resolved before the full potential of nuclear power can be realized. The development of international cooperation in this area may set an example to other spheres, and make more tolerable our existence on this beautiful planet.

REFERENCES

Bickerton, R.J. (1993). "The Purpose, Status and Future of Fusion Research." *Plasma Phys. Control. Fusion* 35, B3-B21 IOP Pub. Ltd.

Carruthers, R.A. (1981). "The Fusion Dilemma." *Interdisciplinary Sci. Rev.* 6 (2): 127–141.

Toschi, R. (1995). "ITER: The World's Fusion Project." *Nuclear Europe Worldscan* 15 (January–February): 55–57.

Williams, L.O. (1994). "The Energy Source: Nuclear Fusion Reactors." *Applied Energy* 47: 147–67.

EXAMPLES AND PROBLEMS

1 Energy generated by fusion reactions

Example: Calculate the energy generated by the reaction of 1 kg of hydrogen by the fusion reaction

$$H + T \rightarrow {}^4He$$

where the atomic masses of hydrogen, tritium, and helium-4 are, respectively, 1.007825, 3.01605, and 4.00260.

Solution: One kilogram of hydrogen would react with $3.01605/1.007825 = 2.99263$ kg of tritium to produce $4.00260/1.007825 = 3.97152$ kg of helium. The mass converted to energy in the reaction is thus:

$$2.99263 + 1 - 3.97152 = 0.02111 \text{ kg} = m$$

The energy generated by this mass conversion is given by Einstein's equation:

$$E = mc^2$$

where c is the velocity of light (2.9979×10^8 m/s). The energy released by the above reaction is thus

$$E = 0.02111(2.9979 \times 10^8)^2$$
$$= 1.897 \times 10^{15} \text{ J}$$

The energy released from this single kilogram would be enough for the whole United Kingdom's electrical energy supply for 2 h (assuming 30% efficiency of conversion). *Problem:* Calculate the energy released by the fusion of 1 kg of deuterium by the fusion reaction

$$D + D \rightarrow {}^{4}He$$

where the atomic mass of deuterium is 2.0140.

2 Design of a Tokamak fusion reactor

Example: Estimate the dimensions and engineering parameters of a possible Tokamak fusion reactor if the limiting heat flux through the wall of the vacuum vessel (first wall) is between 1 and 10 MW/m² and the upper practical limit to the magnetic field is 5 tesla.

Solution: If the magnetic field is B, then in simple terms the magnetic confinement of a Tokamak balances the kinetic pressure of the plasma (proportional to nT, where n is the density and T is the temperature of the plasma) against a magnetic pressure (proportional to B^2). The diagram shows a basic "onion-skin" toroidal reactor. The plasma (of radius a) is surrounded by the vacuum first wall, then the blanket in which the neutrons are absorbed, the energy recovered, and the tritium bred. Next comes a radiation shield and, finally, the superconducting coils to generate the magnetic field.

Given the path length needed to absorb the neutrons and the thickness of shielding required, the thickness of the blanket and shield must be around 2 m. If the area of the first wall is to be large, the plasma radius must be at least the same as the thickness of blanket and shield ($a - t$). Space must be provided for the magnet windings and for access to the center of the torus. Thus, the aspect ratio $A (= R/a)$ should not be less than 3.

This leads to a minimum reactor size with the plasma radius $a = 2$ m and the overall diameter of the torus ~ 20 m. The total area of the vacuum first wall is then 470 m². The power output from the reactor is thus 0.47 GW for a heat flux of 1 MW/m² or 4.7 GW for 10 MW/m².

The Lawson criterion requires that the product of the density of the plasma (n) and the confinement time (τ_p) be sufficient to cause a reaction. A typical target value for

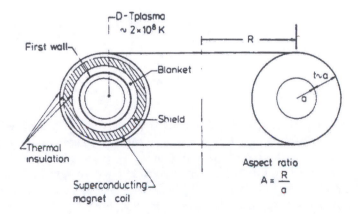

this product would be 3×10^{20} s/m^3, and with a heat flux of 10 MW/m^2, a typical ion density in the plasma would be 3×10^{20} ions/m^3. Thus, in this case the containment time required would be 1 s. It is found that the containment time required is approximately inversely proportional to the heat flux, and therefore the containment time for a heat flux of 1 MW/m^2 would be about 3 s.

As was stated above, the ratio of the plasma pressure to the magnetic pressure (β) is such that βB^2 is related to the plasma density n and, through the Lawson criterion, to the confinement time (τ_p). It is thus found that for a confinement time of 1 s, βB^2 is 2.5, and for a confinement time of 3 s, βB^2 is 0.8. If the magnetic field B is limited to 5 tesla, β must be in the range 0.03–0.1.

The plasma volume is 470 m^3; therefore, the power density varies from 1 to 10 MW/m^3 of plasma volume. This is 10–100 times lower than the power density in the core of a pressurized water reactor (PWR). [With acknowledgment to Carruthers (1981).]

Problem: What would be the consequences for the design of the Tokamak reactor described in the example if the limit on the first wall heat flux were extended to 20 MW/m^2?

3 Lithium cooling of a Tokamak reactor

Example: For the Tokamak reactor described in Example 2 above, lithium is fed to the breeder zone at a temperature (T_i) of 2500C and leaves the zone at $T_o = 600°C$. Calculate the lithium flow rate required and the maximum wall temperature on the lithium side of the first wall if the heat output of the reactor is 2 GW and the heat transfer coefficient (α) between the lithium and the wall is 25,000 W/mK. Assume a specific heat (c_p) for lithium of 3.4 kJ/kg K and a density (ϱ) of 500 kg/m^3. Also assume that 70% of the energy from the reaction that is released to neutrons (80% of the total energy) is converted to heat in the lithium.

Solution: Of the total energy originally generated by the reactor, $(20 + 0.7 \times 80)\% = 76\%$ eventually finds its way into the lithium blanket. The heat released is given by

$$\dot{Q} = V \varrho c_p (T_o - T_i) = 0.76 \times 2 \times 10^9 = 1.52 \times 10^9 \text{ W}$$

where V is the flow rate of the lithium. Thus:

$$\dot{V} = \frac{\dot{Q}}{\varrho c_p (T_o - T_i)}$$

$$= \frac{1.52 \times 10^9}{500 \times 3.4 \times 10^3 \times (600 - 250)}$$

$$= 2.55 \text{ m}^3/\text{s}$$

As explained in Section 9.2, 80% of the energy arising from the fusion reaction is in the form of the kinetic energy of the neutrons, and the neutrons will pass into the lithium, reacting with it to form ^{4}He and T and also releasing heat into the lithium stream. It was assumed that 70% of the original neutron energy (56% of the total energy) is released as heat in the lithium (the remainder being used in the conversion of ^{7}Li; see Section 9.2). Assuming that the remaining 20% of the fusion reaction energy is radiated from the plasma to the first wall, and finds its way into the lithium via that wall, the maximum wall temperature would be

$$T_{max} = 600 + \Delta T \,°C$$

where ΔT is the temperature difference between the wall and the lithium and is given by

$$\Delta T = \frac{2 \times 10^9 \times 0.2}{\text{first wall area} \times \alpha}$$

$$= \frac{2 \times 10^9 \times 0.2}{470 \times 25000} = 34°C$$

Thus, the maximum first wall temperature would be 634°C.

Problem: Repeat the calculations in the example for the reactor calculated on the more relaxed energy flux constraint given in Problem 2.

BIBLIOGRAPHY

American Nuclear Society (ANS) (1983). *Proceedings of the Fifth Topical Meeting on the Technology of Fusion Energy*, Knoxville, Tenn., April 26–28.

Carruthers, R. (1977). "The Fusion Dilemma." *Interdisciplinary Sci. Rev.* 6 (2): 127–41, 1981.(See also *VIII Fusion Prague 1977, 8th European Conference on Controlled Fusion and Plasma Physics*, vol. 2, 217–29.)

Gibson, A. (1977). "The JET Project." *Atom* (254): 3–15.

International Atomic Energy Agency (1982). *Plasma Physics and Controlled Nuclear Fusion 1982, Conference Proceedings*, Baltimore, Md. September 1–8, vols. 2 and 3. IAEA, Vienna.

Lehnert (1977). "Thermonuclear Fusion Power." *Energy Res.* 1, 5–25.

Lomer, W.M. (1983). "Remaining Steps towards Fusion Power." *Nucl. Energy*, 153–57.

Pease, R.S. (1977). "Potential of Controlled Nuclear Fusion." *Contemp. Phys.* 18, 113–35. (See also "Physics in Technology," 144–51.)

——(1978). "The Development of Controlled Nuclear Fusion." *Atomic Energy Rev.* 16 (3): 519–46.

——(1979). "Nuclear Fusion: The Development of Magnetic Confinement Research." *Fusion Technol.* United Kingdom Atomic Energy Authority.

Pease, R.S., and A. Schluter (1976). "The Potential of Magnetic Confinement as the Basis of a Fusion Reactor." In *Nuclear Energy Maturity, Proceedings of the European Nuclear Conference*, Paris, 91–94, Pergamon, Elmsford, N.Y.

Index